AF344341

Equilibrium
Thermodynamics

Equilibrium Thermodynamics ("Single-Axiom" Approach)

for Engineers and Scientists

Part I

R.W. HAYWOOD

Emeritus Reader in Engineering Thermodynamics
and lately
Fellow of St. John's College
in the
University of Cambridge

KRIEGER PUBLISHING COMPANY
Malabar, Florida
1992

Original Edition 1980
Reprint Edition 1992 with corrections

Printed and Published by
KRIEGER PUBLISHING COMPANY
KRIEGER DRIVE
MALABAR, FLORIDA 32950

Library of Congress Cataloging-in-Publication Data

Haywood, R.W. (Richard Wilson)
    Equilibrium thermodynamics for engineers and scientists / R.W. Haywood.
       p.   cm.
    Bibliography: p.
    Includes index.
    ISBN 0-89464-311-8 (pt. 1 alk. paper). - ISBN 0-89464-380-0 (pt. 2 : alk. paper)
    1. Thermodynamic equilibrium. I. Title.
QC318.T47H39 1990
536'.7-dc20   89-2612
                                                      CIP

10   9   8   7   6   5   4   3   2

Men fear thought more than they fear anything else on earth — more than ruin, more even than death. Thought is subversive and revolutionary, destructive and terrible; thought is merciless to privilege, established institutions, and comfortable habits; thought is anarchic and lawless, indifferent to authority, careless of the well-tried wisdom of the ages.

Bertrand Russell

A new scientific truth does not triumph by convincing its opponents and making them see the light, but rather because its opponents finally die, and a new generation grows up that is familiar with it.

Max Planck

Thought looks into the pit of hell and is not afraid. It sees man, a feeble speck, surrounded by unfathomable depths of silence; yet bears itself proudly, as unmoved as if it were lord of the universe. Thought is great and swift and free, the light of the world, and the chief glory of man.

Bertrand Russell

Contents

PART II   DEVELOPMENT OF BASIC CONCEPTS

Steady-flow Processes

Mixtures of Ideal Gases

Air-vapour Mixtures (Psychrometry)

Combustion (Stoichiometry)

Gas Mixtures *(continued)*

Combustion (Energy Analysis)

Thermodynamic Availability IV
Unrestricted Equilibrium with Environment

Preface To Reprint Edition

In writing my book, it was my hope that it would appeal to both engineers and scientists. That this hope has been realised has been evident from the welcome given to it in both engineering and scientific circles.

The original English edition was given a very handsome review in the October 1981 issue of the Chartered Mechanical Engineer (I. Mech. E., London) by Prof. G. F. C. Rogers of the Mechanical Engineering Department of the University of Bristol. He described it as a very polished and scholarly book which should be read by all teachers of thermodynamics, and one which they would read with enjoyment and excitement, as well as profit.

Reviews from other countries were invariably favourable. Within only 8 months of the book's publication came a suggestion from the USSR for a Russian edition and that was published by MIR in Moscow in 1983. At the time of writing, the most recent praise of the book came in December 1988 from a Soviet author, G. P. Gladyshev of the Institute of Chemical Physics of the USSR Academy of Sciences, who kindly sent me an inscribed copy of his own book, *"Thermodynamics and Macrokinetics of Natural Hierarchical Processes"* ("Nauka", Moscow, 1988). In it, he quotes from mine, paying it a generous compliment. He draws attention to the fact that the work of Hatsopoulos and Keenan had been developed further by myself and says that, together, we had made a highly successful attempt to expound the thermodynamics of equilibrium processes on the basis of a single axiom of stable equilibrium, adding – "Detailed accounts of the subject are to be found in Haywood's brilliant book".

The original English edition of my book was published by John Wiley & Sons Ltd, Chichester, in 1980. In view of the very favourable reception of the book worldwide, of which the above instances are only two examples, the fate of the English edition will come as a surprise to the reader. In 1985, after very protracted correspondence with Messrs. Wiley, I learned that two years earlier, without either consulting or notifying me, large numbers of unbound and paperback copies had been destroyed, effectively taking the book prematurely off the market. At my insistence, the copyright was returned to me in late 1985. In the light of these very upsetting circumstances, I am greatly indebted to Mr. Robert Krieger for his agreement not only to publish this Reprint Edition in two separate volumes, but also to commission from me two companion volumes giving fully worked solutions to all the Problems set herein. These will add greatly to the value of the original work.

The two Parts in which the Reprint Edition is now being published are those into which the original book was divided, namely:

Part I – Basic Concepts
Part II – Development of Basic Concepts

They are fully complementary to each other, but issuing them in separate volumes will allow students to make use of Part I in their earlier academic semesters without having simultaneously to carry around the more advanced Part II.

In this Reprint Edition, I have taken the opportunity of adding to the title of the book the words – "*Single-Axiom*" Approach. That approach was described in my original Preface as the most important of the recent developments in equilibrium thermodynamics, and that fact has seemed to me to warrant this addition to the title.

The new approach followed in this book, and in the earlier book by Hatsopoulos and Keenan[1], was first given the title "Single Axiom" in their paper [a] of 1962. This was followed by the publication of their book in 1965. The importance of that approach lies not in the question of a "single axiom" but in the starting point from which it sets out, the very important Law of Stable Equilibrium, from which the previously disparate so-called First and Second "Laws" are deduced as corollaries. This is depicted clearly in the thermodynamic "family tree" displayed below, taken from a paper [b] which I presented at a Teaching Workshop in Thermodynamics in Cambridge, England in 1984. Readers may also find of interest a paper on a similar topic which was presented at the Teaching Workshop by P. H. Brazier [c], though they should note that the systems about which he writes should have been defined as *Constrained Systems*.

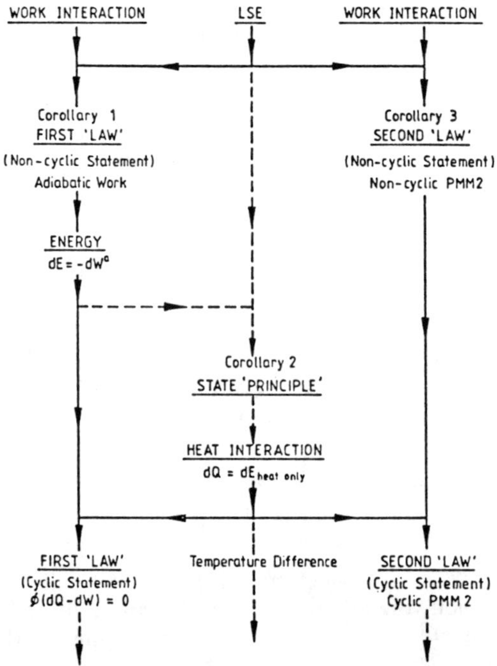

In the context of both this book and that by Hatsopoulos and Keenan, it is quite immaterial as to whether or not the new approach can be legitimately claimed as being based upon a single axiom. That claim has been challenged ever since Hatsopoulos and Keenan made it in the title of their paper. This question was again raised during the discussion at the Teaching Workshop in Cambridge. To that questioning I replied that, because I considered the *so-called* Single-Axiom Approach to be a method of presenting classical thermodynamics that was so much more logical than, and so greatly superior to, the more conventional 'Keenan' approach of 1941, I was not prepared to throw away the baby with the bathwater simply because some raised doubts as to whether it was, in fact, based on a single axiom. I was, and still am, happy to let the philosophers argue at length about that point. *The Single-Axiom Approach* has become a conveniently short label of identification, to distinguish it from the less satisfactory approach followed earlier by Keenan and many others. It is for that reason that I have incorporated the term in the title of this Reprint Edition, though with the words ''Single-Axiom'' placed within inverted commas in order to indicate the possible uncertainty as to their validity. Again, that does not indicate any uncertainty as to the validity of the mode of presentation of classical thermodynamics followed herein.

In this Reprint Edition of my book, care has been taken to eradicate errors which appeared in the original edition. The majority of those were due to lack of care by the printers, but there was a serious error of my own making in paragraph (c) of Section 9.4, and the accompanying Fig. 9.1. I apologise to any earlier readers who may have been misled by this. It was put right in the Russian edition and has also been put right in this. The original error arose from the fact that, in an early draft for the original edition, I followed Hatsopoulos and Keenan in the definition of a Cyclic PMM 2 in their Section 13.4. There they defined a Cyclic PMM 2 as 'any heat engine' (which I prefer to call a Cyclic Heat Power Plant or CHPP) 'which exchanges heat *with a single thermodynamic system in a stable state* and delivers net work'. I later came to the conclusion that it was not necessary to include within the definition of a Cyclic PMM 2 any reference to a thermodynamic system in a stable state. Accordingly, that was omitted in the final version of my definition of a Cyclic PMM 2 in Section 8.6; unhappily, however, I failed to notice that I had not similarly rewritten my earlier draft of paragraph (c) of Section 9.4, nor altered Fig. 9.1(c) accordingly. I explain this in some detail here, since it is a point of some technical importance.

I cannot end this Preface without a very warm expression of thanks to my former colleague at Cambridge, and long-time collaborator in the writing of technical papers, Dr. John H. Horlock, Vice-Chancellor of the Open University in the UK. Sharing my belief that the book, recognised world-wide as a scholarly work, should be rescued from the fate so peremptorily assigned to it by the original publisher, it was he who kindly put me in touch with Mr. Robert Krieger. Many will share my gratitude to them both.

[a] Hatsopoulos, G. N. and Keenan, J. H., A single axiom for classical thermodynamics, *J. App. Mech.*, 29, 193–199, 1962.

[b] Haywood, R. W., Teaching thermodynamics to first-year students by the Single-Axiom Approach, *Teaching Thermodynamics*, ed. J. D. Lewins, 205–216, Plenum Press, New York, 1985

[c] Brazier, P. H., The Thermodynamics Laws from the Law of Stable Equilibrium, *ibid*, 217–227.

R. W. HAYWOOD

Preface

Equilibrium thermodynamics is basically concerned with the *macroscopic* (gross) behaviour of systems that are subject to processes which take them from one state of stable equilibrium to another while the system may interact energy-wise with its environment. It treats the material within the system as a continuum by ignoring the particulate nature of matter and the quantization of energy. These are the concern of *statistical thermodynamics* in its prediction of macroscopic behaviour through the study of events occurring at the *microscopic* level. Thus, equilibrium thermodynamics, with which this book is principally concerned, constitutes essentially a study of the relations between work, heat, and the properties of systems; it is this which makes it of great importance to the engineer, particularly in the field of energy conversion. Not only must the engineer devise *performance parameters* by which to express the actual performance of his work-producing and work-absorbing devices, but he must also be able to establish *performance criteria* against which to judge the actual measured performance. It is the science of thermodynamics which enables him to do this on a rational basis.

Through the action of effects which for the present we may loosely term 'frictional' or 'dissipative', all natural processes are in some degree thermodynamically imperfect, termed by the thermodynamicist 'irreversible'. Consequently, either less work is produced or more work absorbed in real-life plant than would be the case in that imaginary and idyllic land of the thermodynamicist which we may call 'Thermotopia'. Here all processes are, in his language, 'reversible', so that there are no such imperfections. In this context, it is worth making the point that, as do all other scientists, thermodynamicists construct a simplified *model* to describe the behaviour of the physical world, that behaviour being of too complicated a nature to allow every facet to be taken into account at a first encounter.

Because all natural processes are in some degree imperfect (irreversible), the engineer cannot use the results of experiments to enable him to set up his performance criteria by which to judge the degree of excellence of his plant. Instead, he is only able to call upon the resources of the human intellect through the rational application of scientific thought. In this, the science of thermodynamics provides an outstanding example of the power of abstract thought.

The work of the engineering thermodynamicist is not ended when he has devised expressions, in the shape of algebraic formulae, for the respective performance parameters and criteria appropriate to each of his devices. He will then need numerical data on the thermodynamic properties of the working substances which are either contained within or passing through these devices. Here he will need to call upon both experimental and mathematical expertise. From this activity may come empirical or semi-empirical correlating formulae to fit the great wealth of experimental data. However, the values of the many thermodynamic properties of a given substance are all theoretically interrelated and in the revelation of these interrelations lies one of the great achievements of the science of thermodynamics. The mathematical equations which have been devised to fit the experimental data will therefore have to be put together in a form that will maintain thermodynamic consistency between the numerical values computed therefrom by the engineer. This task will fall upon the thermodynamicist.

The considerations outlined above have been influential in shaping the layout of this book, making it a volume designed for the engineer rather than for the physicist or chemist, though the latter will also have much to learn from it. However, other influences have been equally strong in giving it a distinctive character. In spite of the fact that there have been important developments in recent years, the subject matter of thermodynamics is still largely taught along lines which follow closely those set out by the great pioneers of the last century, who were at the time exploring the unknown. This book takes advantage of recent developments to restructure the presentation into a more logically satisfying sequence.

The most important of the recent developments in equilibrium thermodynamics has been the *Single-Axiom Approach* of Hatsopoulos and Keenan,[1] in which the previously disparate 'Laws' of thermodynamics have been shown to follow logically from a single basic *Law of Stable Equilibrium*. The next most important development is associated with the topic of *thermodynamic availability* and the associated concept of *exergy*. The topic of thermodynamic availability deals essentially with the availability of energy for work production, a factor which has become of increasing importance in very recent times as a result of the emphasis on energy 'saving'. Despite the fact that this topic originated with Willard Gibbs and Clark Maxwell over a century ago and has been pursued with some vigour in Germany, it has in general suffered comparative neglect, although the respective theorems have recently been formulated concisely in a critical review[2] by the present author. This book takes advantage of both these advances in our understanding. The numerous concepts and theorems of equilibrium thermodynamics are thereby restructured and presented in a more logical sequence than hitherto.

It is the author's experience that the important advances made by Hatsopoulos and Keenan have not received the recognition which they deserve in academic circles because the material appears in a volume which is not only set at a level of sophistication beyond the normal capacity of the undergraduate but also goes well beyond the content appropriate to an undergraduate course. The basic ideas, however, are not of great difficulty and the present author has always believed that they could be presented in a form which, while remaining equally rigorous, was set at a level which would be readily digestible by an undergraduate of reasonable competence coming fresh to a study of the science of thermodynamics. He has been confirmed in this belief by giving a short course of undergraduate lectures in which, over a number of years, he has been able to develop a mode of presentation of the ideas of Hatsopoulos and Keenan in more readily assimilable form. The interest displayed by the undergraduates has been very rewarding and the present volume is the natural outcome of this venture.

An earlier book[3] published by Keenan alone in 1941 had a greatly beneficial influence on the teaching of thermodynamics in schools of engineering in the United States and the United Kingdom. However, because the book developed the concepts and theorems of classical thermodynamics from *cyclic* statements of the so-called First and Second Laws, it had the unfortunate effect of focusing undue attention on cyclic processes at the expense of non-cyclic processes, which are the more natural. By contrast, the Law of Stable Equilibrium of Hatsopoulos and Keenan, from which those 'Laws' follow as corollaries, relates essentially to *non-cyclic* processes, as also do the theorems of thermodynamic availability. The cyclic approach unfortunately conceals for too long the nature of the true source of irreversibility, but the non-cyclic approach brings this out very clearly right from the start. Moreover, cyclic processes are somewhat artificial constructs. The naturally occurring processes of the physical world are basically non-cyclic in character, a cyclic process merely constituting the special case in which, by a deliberately created succession of non-cyclic processes, the final thermodynamic state of a system is made to coincide with its initial state. Furthermore, if we start with unproven propositions relating to cyclic processes, we are not led naturally into the theorems of thermodynamic availability, which relate to systems that are carried by non-cyclic processes between specified end states in the presence of a specified environment. On the other hand, when we take advantage of the Single-Axiom Approach of Hatsopoulos and Keenan to develop the concepts and theorems of equilibrium thermodynamics in terms first of non-cyclic processes, these important availability theorems take an earlier and more natural place in the sequence of ideas. Not only is there no longer any need to introduce the troublesome *Clausius Inequality* as a preliminary to the establishment of the concept of entropy (that Inequality itself essentially constituting a theorem in thermodynamic availability), but the whole sequence of ideas then follows a more logical pattern. That pattern is followed herein.

As the concepts and theorems are developed, chapter by chapter, from the starting point of the basic Law of Stable Equilibrium, the reader will find that the propositions which have been given the titles of the First and Second Laws of Thermodynamics take on the nature of corollaries, so ceasing to be basic 'Laws' in their own right; moreover, no need is found for the so-called 'Zeroth Law'. In order to help the reader to follow the logical development of the long sequence of ideas that form the groundwork of the science of thermodynamics, a 'Family Tree of Thermodynamics' is constructed and developed in the earlier chapters, its step-by-step growth being shown at the end of each relevant chapter. By this means, the logical structure of this rather abstract branch of science is made clearly evident.

The Second Law of Thermodynamics is frequently, but quite undeservedly, endowed with a certain mystical aura, which should now be firmly dispelled by its relegation to a status subsidiary to that of the Law of Stable Equilibrium, of which it is merely a corollary. No other important branch of science has been so dependent on so many unproven postulates as are represented by the so-called Zeroth, First, and Second Laws. It must be a matter of satisfaction that the science of thermodynamics now needs no such underpinning.

Close on fifty years ago, Keenan described the steady-flow availability equation as promising to be as revolutionary in its effect on thermodynamic reasoning as had been the development of the steady-flow energy equation in its time. Both equations relate to non-cyclic processes. That the importance of the concepts and theorems of availability have not even yet taken the position of importance that is undoubtedly their due must almost certainly be attributable to the fashion of starting the presentation of thermodynamics in terms of cyclic statements of the First and Second Laws, rather than starting with the study of non-cyclic processes and proceeding therefrom to cyclic processes, as in this book.

Textbooks on thermodynamics come on the market in a steady stream, most of them little different from each other. There would consequently be little justification for writing yet another if it did not have something new and important to say. However, that something must be expressed in terms which are readily assimilable by the undergraduate if the new approach is to stand a chance of gaining widespread acceptance. It is considered that this book is set at such a level and that it will help to bring thermodynamics out of its nineteenth-century setting, in which it has remained for too long. It is, indeed, unthinkable to the author that the recent developments of which it has taken advantage should continue to be neglected. It is a matter of some concern that, in an age in which the great importance of energy 'saving' is increasingly being appreciated, important books and papers on energy 'conservation' are still appearing in which the concept of the availability of energy for work production is found not to merit even a mention.

The book is in two parts, the first dealing with basic concepts and the second with the development of those concepts. The last two chapters of Part II give a fairly extensive and considerably more careful treatment of chemical thermodynamics than is found in most texts. These should appeal particularly both to the power engineer concerned with combustion and other chemical processes and to the chemical engineer.

The author's *Thermodynamic Tables in SI (metric) Units*[4] have been used in calculating the answers to the extensive collections of Problems set at the end of Part I and the end of Part II.

This book is the product of many years of teaching experience and owes much to the help received from many people, not least from those whom the author has taught, and, indirectly by their writings, from the late Professor Joseph H. Keenan and his collaborators. The author is also particularly indebted to his colleague Dr Martin D. Cowley, of Trinity College, for reading through early drafts of the manuscript and for his many helpful and perceptive comments. Finally, a warm word of thanks is due to Miss Jill Stroud for her patience and skill in the typing of the manuscript.

Cambridge, England   RICHARD WILSON HAYWOOD

REFERENCES

1. Hatsopoulos, G. N., and Keenan, J. H., *Principles of General Thermodynamics,* John Wiley & Sons Inc., New York, 1965.
2. Haywood, R. W., A critical review of the theorems of thermodynamic availability, with concise formulations: Part 1, Availability; Part 2, Irreversibility, *Jl. Mech. Engg. Sci.,* **16**, Nos. 3 and 4, 160–173, 258–267 (1974) and **17**, No. 3, 180 (1975).
3. Keenan, J. H., *Thermodynamics,* John Wiley & Sons Inc., New York, 1941.
4. Haywood, R. W., *Thermodynamic Tables in SI (metric) Units,* with enthalpy–entropy diagram for steam and pressure–enthalpy diagram for refrigerant-12, 2nd ed., Cambridge University Press, 1972.
   Also in Spanish translation – *Tablas de Termodinámica en Unidades SI (métricas),* (Trans. by A. F. Estrada), Compañia Editorial Continental, S.A., México, 1977.

PART I

BASIC CONCEPTS

New times demand new measures and new men;
The world advances and in time outgrows
The laws that in our father's day were best;
And doubtless, after us, some purer scheme
Will be shaped out by wiser men than we,
Made wiser by the steady growth of truth.

James Russell Lowell

CHAPTER 1
Some Thermodynamic Terms

In this chapter, as an introduction to our study of equilibrium thermodynamics, we present formal definitions of some of the terms which we shall encounter during the course of our studies. We then follow the presentation of these definitions with a preliminary discussion of the concepts of energy, work, heat, and temperature. Formal definitions of these latter terms will be presented in subsequent chapters.

1.1 System

A *system* is any collection of matter contained within prescribed boundaries which may move but across which no matter is able to pass.

The fluid contained within the cylinder and leak-tight piston of Fig. 1.1 is such a system.

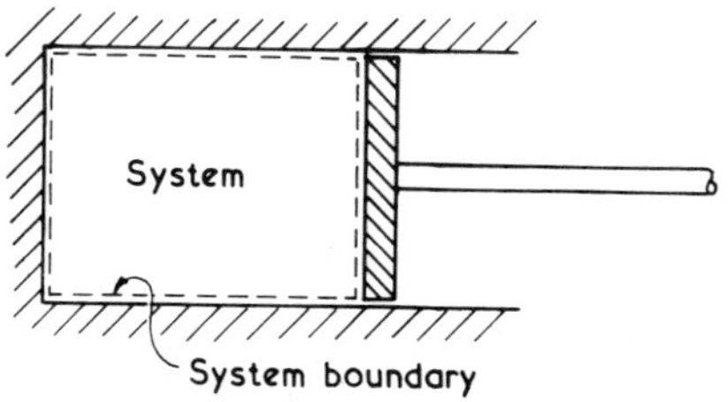

Fig. 1.1 Definition of a system

1.2 Simple System

A *Simple System* is any system which is macroscopically homogeneous and isotropic* and whose *internal* condition is negligibly influenced by the effects of surface tension (capillarity), external force fields (electric, magnetic, and gravitational), and the distortion of solid phases (shear).

---

* A system is said to be *isotropic* when its physical properties do not vary with direction, so that there are no preferred directions within the body of the system.

1.3 Neutral property

A *neutral property* of a system is one whose changes in value can be achieved by only *temporary* changes in the environment. (For example, the *position* of a marble on a smooth, horizontal plane surface is a neutral property of the marble. A change in position can be brought about without leaving any finite trace in the environment.)

1.4 Thermodynamic property

A *thermodynamic property* is, per contra, one whose change in value *cannot* be brought about simply by temporary changes in the environment. (For example, *volume* and *pressure* are thermodynamic properties.)

1.5 Extensive property

An *extensive property* is one for which the value of the property for the entire system is equal to the sum of the values of that property for all elements of the system (namely all subsystems into which the system may be subdivided). Examples are mass, volume, and energy; we shall encounter others as our studies in thermodynamics proceed.

1.6 Intensive property

An *intensive property* of a system is one which would not approach zero for systems of progressively smaller size and tending to a point. Thus an intensive property has a finite value at a point. Examples are pressure, temperature, and specific volume (volume per unit mass); mass and volume themselves clearly do not meet the definition of an intensive property.

1.7 State

The *state* of a system is its condition at any instant, as identified by the values of its primitive properties, namely those (such as position, volume, pressure, etc.) whose values can be determined without changing the state of the system.

1.8 Thermodynamic state

The *thermodynamic state* of a system is the condition of a system as identified by the values of its thermodynamic properties. (In our thermodynamic studies, for the sake of brevity, we shall frequently refer to this as simply the *state* of the system.)

1.9 Intensive state

The term *intensive state* refers to the *local* state of an *element* of the system and is identified by the values of the intensive properties of that element. For a nonuniform state of a system, the intensive state would vary from element to element. When the value of each of the intensive properties of a system is the same in every part of the system, the latter is of uniform intensive state throughout.

1.10 Phase

A *phase* is here defined as the collection of all *homogeneous* parts of a system that have the same intensive state.

Thus a mixture of solid, liquid, and vapour coexisting together within a single system is a mixture of these three phases. Each of these phases is homogeneous and of the same intensive state throughout itself, but has an intensive state different from that of the other two phases, which are of different chemical aggregation.

1.11 Process

The *process* involved in a given change of state of a system is described in terms both of the circumstances which bring about that change of state and of the *interactions* that occur between the system and its environment during the change of state. Such interactions may, for example, take the form of *work* and *heat*.

1.12 Path

The succession of states through which a system passes when subject to a given process is described as the *path* traced out by the system during that process. In general, only for relatively slow (*quasistatic*, Section 2.13) changes of state is it possible to identify the path.

1.13 Cyclic process

A system undergoes a *cyclic process* when the end state of the system is identical to the initial state; the system is then said to undergo a *thermodynamic cycle*.

Attention is drawn to the note in Section 7.2 pointing out that in classical thermodynamics, which treats matter as a continuum, it is only meaningful to deal with cyclic processes which start from either a Stable State or a quasistatic Stable State (Section 2.13).

1.14 Thermodynamic property change during non-cyclic and cyclic processes

The following important consequences follow from the foregoing definitions:

(a) The change in value of any thermodynamic property P during any process undergone by a system between specified states is dependent only on those end states of the system (i.e. it is independent of the nature or path of the process by which the change of state is brought about).

6

(b) For a system that has undergone a completed cyclic process,

$$(\Delta P)_{\text{cycle}} = 0, \tag{1.1}$$

or, if the path can be identified,

$$\oint dP = 0, \tag{1.2}$$

where the symbol $\oint$ denotes integration round the cyclic path.

1.15 A preliminary discussion of the concepts of energy, work, heat, and temperature

Some notions of these four concepts will already be held in the mind of the reader, who nevertheless would probably find it quite taxing to define them concisely and precisely. This would be particularly the case in relation to the concept of heat, which many laymen erroneously treat as if it were resident *in* a body. We shall in due course present careful definitions of each of these concepts, but we would find it extremely inconvenient if we were not allowed to use these terms until we had done so. Consequently, while using them at this preliminary stage we shall be taking advantage of the reader's existing body of knowledge. At the same time, it is necessary to warn that this may be imperfect, or even erroneous.

1.15.1 Energy

Since we discuss *energy* first, it is necessary to point out that its formal definition in Chapter 5 will follow the formal definition of *work* in Chapter 3.

Energy has been loosely defined as 'the capacity to produce an effect'. However, this is not acceptable as a scientific definition, although the reader will recognize that it has some validity in relation to the circumstances associated with the use of such terms as potential energy, kinetic energy, thermal energy (as distinct from heat), and nuclear energy. These are forms of energy which a system may possess and it is clear that we need a precise definition which will enable us to include all of these manifestations within the terms of the thermodynamic property that we call energy.

We know that energy may not only reside in a system but may transfer from one system to another. Such *energy transfer* is a transitory phenomenon, in that it is associated with interactions between systems and persists only while the interaction persists. Work, heat, and electromagnetic radiation are different modes of energy transfer and, as already mentioned, we shall indeed define work before defining energy.

1.15.2 Work, adiabatic walls, and adiabatic processes

The form of work which will come most readily to the mind of the reader is the *mechanical work* performed by an agent exerting a force on a body and moving that body in the direction of the force. However, as we shall formally establish, there are other forms of energy transfer which, for our purposes, can also be recognized as work. Such are the *electrical work* supplied by a battery to a motor which is used to raise a weight and the *magnetic work* represented by the energy transfer between the two sides of an electric transformer. It is again clear that we need a precise definition which will cover all forms of work transfer.

Work appears in consequence of an interaction between systems, and we find it convenient to coin a word which describes a 'work-only' situation; this is the word *adiabatic*. Thus, if the boundary of a system is such that it permits *only work interactions* with the environment, we describe the boundary as an *adiabatic wall* and the processes which the system undergoes while so restricted as *adiabatic processes*.

Another possible type of interaction between systems is a *heat interaction*, which we shall discuss next. We mention it here in order to warn the reader that the word *adiabatic* is frequently used to imply a 'no-heat-transfer' situation. We choose, however, to use it in the more precise and restricted sense defined above, namely in relation to a 'work-only' situation.

1.15.3 Heat and temperature

The term *heat* is more frequently misused by the layman, and indeed by many scientists and engineers, than any other term encountered in the science of thermodynamics. Too frequently do we find the use of phrases such as 'the heat in the sea', 'the heat possessed by a hot body', etc., when reference is in fact being made to the energy content of a body and not to heat. As we have already noted in Section 1.15.1, heat is a mode of *energy transfer* which persists only while the heat interaction persists. We shall need to give a formal definition for this particular mode of interaction between systems and shall do so in Chapter 6, after we have first defined work and then energy. In the meantime, we will make do with a looser understanding of the term which will here meet our immediate requirements. At this level, we recognize that, if we bring a mercury-in-glass thermometer into contact with a body which feels warmer to the touch than another, the mercury will rise to a higher level than when the thermometer is brought into contact with the cooler body. The height of the mercury serves as an arbitrary indicator of what we call the *temperature* of a body; for temperatures deduced from this arbitrary type of temperature indication we shall use the symbol θ. However, we shall later need a more precise and scientific definition of this property of a body. We shall provide this by defining in Chapter 11 a property called *thermodynamic temperature*, for which we shall use the symbol T. Nevertheless, the former arbitrary concept of temperature will serve us for the time being. We must point out, however, that an instrument such as a mercury thermometer in fact only enables us to detect that there may be a *temperature difference* between two bodies, in that, if the thermometer records different levels when placed in contact with two bodies in succession, this denotes a difference in temperature between the two bodies. Equality of mercury level denotes equality of temperature, but the mercury level gives us no idea of the 'absolute' temperature of a body.

8

When we bring two solid bodies of different temperature into what we may call 'thermal contact', we find that the hotter body is cooled and the cooler body is warmed until a state of mutual equilibrium between the bodies is reached. When a mercury thermometer is then brought into contact with the two bodies in turn, the mercury will be found to settle to the same level on both occasions. In this process of reaching equality of temperature, we say that the hotter body has transferred some of its *energy* to the cooler body and, *while in transit*, this energy transfer is described as *heat*. We note particularly that it was not heat, but energy, that resided in either body and that heat is simply *energy in transit* as a result of this particular mode of interaction between bodies. In other words, heat, like work, is a transitory form of energy transfer, persisting only while the interaction persists.

We conclude from the above description that the 'motive force' of heat transfer is a difference in temperature. We might therefore be tempted at this stage to define heat as 'energy in transit as the result of a temperature difference', as many authors have in effect done. However, this would be putting the cart before the horse, for we shall first define a heat interaction in Section 6.2 and then, in Section 6.6, define temperature difference in terms of a heat interaction.

1.15.4 Distinction between work and heat

Since we find that work and heat are both modes of energy transfer, it might appear that there is no great need to make a clear-cut distinction between them. However, our thermodynamic studies will show that there are circumstances in which, from a scientific point of view, it is very important indeed to make such a distinction, even though in practice it may sometimes be virtually impossible to separate one type of interaction from the other. Such a difficulty arises when friction plays a prominent part. Thus, when we push a brick along a rough surface we are clearly doing work on the brick; at the same time, we find that warmth spreads through both the brick and the surface, indicating that the mode of energy transfer known as heat is occurring within both. We are then curious about events occurring at the frictional interface. There we are clearly in difficulty if we attempt to separate the two types of interaction by some discrete boundary. In practice, engineers and scientists can only handle their calculations of work and heat flow in situations in which there is a clear separation of identity. However, they sometimes idealize the actual situation somewhat in order to achieve this separation, if only in the mind.

1.16 Summary

In this chapter we have been preparing the ground for our formal studies in equilibrium thermodynamics. We have done this by defining terms which have a particular meaning or significance in this branch of science and then by engaging in a preliminary discussion of the concepts of energy, work, heat, and temperature. We shall formally define these in later chapters in the sequence work, energy, heat, and temperature difference, ultimately defining a scientific measure of temperature in Chapter 11.

CHAPTER 2
Thermodynamics and Equilibrium

2.1 The nature of thermodynamics

The science of *thermodynamics* studies the *states* and *properties* of physical *systems* and the changes in these states and properties resulting from *processes* which the systems undergo.

In general, during the course of a process there will be *interactions* between a system and its *environment*, but there is an important class of process during which a system may undergo a change of state even though completely isolated from all interaction with its environment; such is the process that occurs when a system in isolation settles from a *non-equilibrium state* to a final unchanging *thermodynamic state of stable equilibrium* which, for the sake of brevity, we call simply a *Stable State*. As a simple example, we may cite the case of a stirred fluid which is then isolated. In consequence of the viscosity of the fluid, the eddies created during the stirring process are destroyed by frictional dissipation and the fluid eventually settles to an unchanging *macroscopic* Stable State (though, on a *microscopic* scale, random kinetic motion of the individual molecules continues).

Equilibrium classical thermodynamics is concerned essentially with processes between *Stable States*, because it deals only with phenomena occurring on a macroscopic scale; i.e. it treats as a *continuum* the matter contained within any system, such as that of Fig. 1.1. This inevitably sets a limit to one's understanding of the behaviour of physical systems. On the other hand, by starting with events occurring at the microscopic level, *statistical thermodynamics* takes account of the *particulate* nature of matter and the *quantization* of energy, thence building up predictions of the macroscopic behaviour of systems. One's understanding of the behaviour of physical systems is thereby enhanced. However, this book is confined to a study of the fundamentals of equilibrium thermodynamics, with particular reference to engineering processes. As an introduction to this study, we first discuss and define certain terms which will frequently recur and which bear a special connotation in thermodynamics. Before we can discuss the behaviour of a system in scientific terms, we must not only define the system itself but also identify the constraints that limit the range of states that the system can take up.

2.2 Interactions

We have already noted that a change of state of a system may result from interactions between a system and its environment, though it can also occur in the

10

absence of such interactions if, internally, the system is initially in a non-equilibrium state.

In our study of thermodynamics, an interaction between systems may take the form of either a *work interaction* or a *heat interaction*, though both types of inter-action may occur simultaneously. Both types involve a *transitory* mode of energy transfer.

2.3 Isolated systems

An *isolated system* is one which is cut off from all possible interactions with its environment, namely with other systems. To this extent it is somewhat of an idealization, to which real systems can approach in varying degrees, depending on the individual circumstances.

2.4 The possible states of a system – Constraints and Allowed States

The range of different states which it is theoretically possible for a given system to assume is, in general, very wide indeed. Under specified conditions, however, the *constraints* to which the system is subject allow it to take up (by interaction with its environment or otherwise) only certain of the totality of states which would otherwise be accessible to it. These constraints may be both internal and external to the system, the latter being imposed on the system by the environment.

By way of introducing the terms and concepts with which we shall be dealing, we consider first a few very simple systems. We shall then be able to treat the behaviour of systems in general terms.

2.4.1 Simple pendulum in an air-filled box

Figure 2.1 depicts a simple pendulum comprising a bob suspended by a light inextensible cord in the Earth's gravitational field of intensity g. The pendulum is mounted in a rigid box containing air. We shall define as our system the entire con-tents of the box lying within the depicted boundary. We shall examine the con-straints to which this system is subject and we shall describe as *Allowed States* of the system all those states which the system is potentially capable of taking up, by interaction with the environment or otherwise, while it remains subject to these constraints.

Considering first simply the mechanics of the pendulum alone, it is clearly evident that the inextensible cord provides a constraint internal to the defined system; no hypothetical state in which the bob were at a radial distance from the point of suspension greater than r could be among the Allowed States of the defined system while the latter remained subject to this *internal constraint*.

The presence of the Earth's gravitational field also exercises a constraint on the system, in that the pendulum can take up only those states of simultaneous position and velocity which, in accordance with the laws of mechanics, are con-sonant with its presence in this field. In particular, we may note that there is only

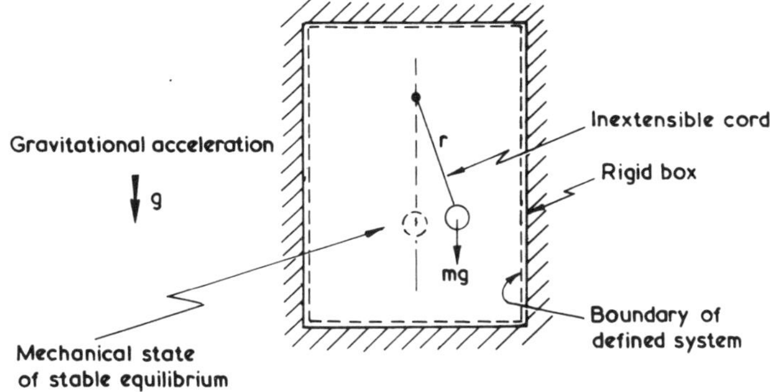

Fig. 2.1 Simple pendulum in an air-filled box

one position in which the pendulum, after initial displacement, will ultimately come to rest by virtue of viscous damping in its passage through the air. This rest position, with the bob vertically below the point of suspension, is the one and only *mechanical* state of stable equilibrium that the pendulum can take up while it is constrained by the given gravitational field. This field consequently acts as an *external constraint* on the defined system.

Instead of considering simply the mechanics of the pendulum alone, we can now widen our horizon by applying our knowledge of both mechanics and thermodynamics to the entire system comprising both the pendulum and the air contained within the depicted boundary. This system is subject to the two constraints already discussed, together with one further external constraint, namely that restriction imposed on its *volume* by the rigid walls of the containing vessel; in no Allowed State of the system can the volume be greater than this specified value. The Allowed States of this more extensive system are of immeasurably greater range than those of the pendulum alone. Furthermore, instead of only one *mechanical* state of stable equilibrium for the pendulum, there is an immeasurably large number of possible *thermodynamic* Stable States of the complete system, namely one for each possible state of the air when it and the pendulum are at rest. To take the system from one of these Allowed Stable States to another, one would only need to allow the system to settle to a new Stable State after letting it exchange heat temporarily with its environment while the three constraints on the system remained unaltered.

In considering the possible Stable States of the complete system, we now come to a most important observation. Let us imagine that the pendulum is held initially in some position displaced from the vertical, with the air still and in a certain state (i.e. at a certain temperature and pressure). Let us then suppose that, after release of the pendulum, the system is completely isolated from all interaction (i.e. work-wise and heat-wise) with its environment. To achieve such complete isolation we would need to surround the system by a hypothetical perfect thermal insulator, forming what is commonly called an *adiabatic wall*.* We have no such thing in prac-

---

* A more specific definition of this term is given in Section 3.6.

12

tice as a perfect thermal insulator, but we can obtain a close approach to it. If we succeeded in obtaining perfect isolation, we would find that, as a result of viscous dissipation, the pendulum would eventually come to rest in its stable position and all eddies in the air would eventually subside, leaving the air in an unchanging Stable State at slightly higher values of temperature and pressure than obtained initially. (Note that the Earth's gravitational field does *not* do work on the pendulum as the latter descends; in descending, its decrease in potential energy is transformed into kinetic energy and this kinetic energy is gradually completely dissipated by frictional dissipation as the pendulum moves through the air, whose energy is thereby increased. We have, of course, yet to define *energy*; this will be done in Chapter 5.) Our most important observation is that, however many times we repeated the experiment, we would always find that, from the same initial state, the system would always settle to this same final Stable State when completely isolated from its environment, *provided that all three constraints on the system remained unaltered* (e.g., provided the cord did not break).

From such an experiment, it would thus appear that, *if none of the constraints on a thermodynamic system are altered, it will not only always settle in time to a Stable State when completely isolated from its environment but, from any given initial state, will always settle in these circumstances to one and the same Stable State.*

From this single example, we have, of course, no right to cite the above statement as a general conclusion; nor indeed can we find any proof of its applicability to constrained systems in general. However, we shall in due course enshrine such a statement in a general proposition called the Law of Stable Equilibrium. The importance of this general proposition lies in the fact that from it we shall be able to develop, in logical sequence, all the theorems of classical equilibrium thermodynamics, including the so-called First and Second Laws. We have first, however, to continue our preliminary study of a few very simple systems in order to familiarize ourselves with some of the terms and concepts which will frequently recur in our work.

2.4.2 Compound pendulum in an air-filled box

We next replace the simple pendulum by a compound pendulum, in which the bob is secured to a rigid rod, as depicted in Fig. 2.2. This sytem is still subject to one internal and two external constraints, in that (a) oscillation of the bob takes place at a fixed radius about a fixed axis, (b) there is an externally imposed gravitational field, and (c) the system is confined within rigid walls.

Apart from the motion of the bob now being confined to a fixed circle, so reducing the range of Allowed States, there is only one important difference between this and the previous case: there are now two positions in which the pendulum could remain at rest, namely positions (1) and (2) in Fig. 2.2. Both are therefore equilibrium positions of a kind, but only position (1) represents a mechanical state of stable equilibrium. Position (2) is described as a state of *unstable equilibrium*, in that only the merest touch would displace the pendulum

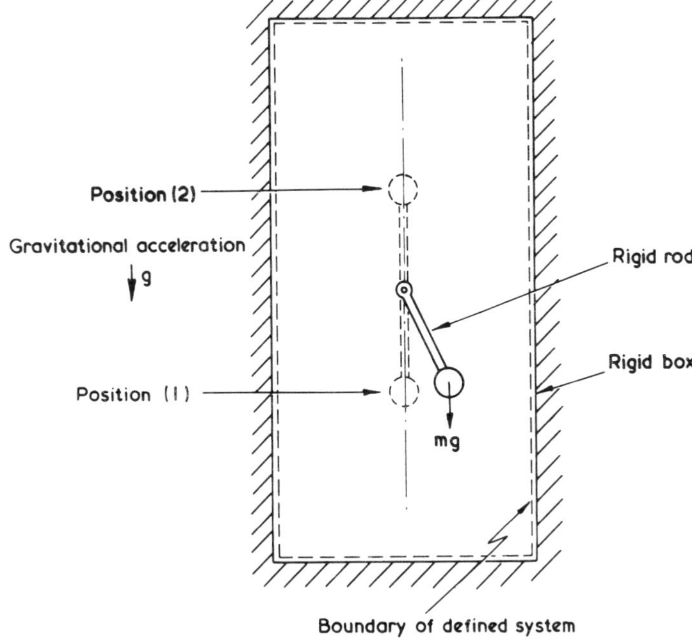

Fig. 2.2 Compound pendulum in an air-filled box

by the infinitesimal amount required to put it into an unstable position, from which it would settle of its own accord to position (1) without the need for any finite interaction with the environment. Hence, as before, we find that, provided that the constraints on the system are not altered, there is one and only one Stable State to which, *from any given initial state*, the defined system would settle after being isolated from all interaction with its environment.

2.4.3 Rigid spherical ball in an air-filled box

Our next simple example is that of a rigid spherical ball in a rigid box containing air, the floor of the box being smooth and horizontal, with the system subject to the Earth's gravitational field. We will suppose that the air is initially still and that, by some means, the ball is initially held up against the roof of the box and is then released, the defined system being thereafter completely isolated from its environment. We would then find that, through viscous damping of the air, the ball would eventually come to rest on the floor of the box, with the defined system in a final Stable State in which the air was at slightly higher values of temperature and pressure than obtained initially. Now, however, there would appear to be many possible alternative Stable States to which the system could thus settle from the given initial state, in that the ball could come to rest at any position on the floor. However, the *position* of a ball on a *smooth horizontal* floor is a *neutral* property of the system (Section 1.3), not a *thermodynamic* property (Section 1.4), so we do not reckon these different positions as different *thermodynamic* states to which the isolated system will settle from a given initial state while the constraints on the system are not altered.

We may note in passing that, not only do all possible positions of the stationary ball on the smooth horizontal floor fall within a single thermodynamic Stable State when the ball is in thermal equilibrium with the air at a final steady temperature, but so also do all possible instantaneous positions and velocities of the air molecules at this *macroscopic* Stable State. Only if we disturb the macroscopic uniformity of temperature and pressure, e.g. by stirring, do we change the state of the air to a non-equilibrium state from which, on isolation, it would settle to a new macroscopic Stable State.

2.4.4 Effect of relaxation of an internal constraint

As an illustration of the effect of the relaxation of an internal constraint on the range of Allowed States of a system, we may consider the system depicted in Fig. 2.3. Here a rigid box is initially divided by a rigid diaphragm, with two inert gases A and B initially at different Stable States on either side of the diaphragm and with the pressure p_A greater than p_B. The diaphragm clearly acts as an internal constraint on the system, in that, while it is intact, no state in which some of gas A is inside side B can be an Allowed State of the system.

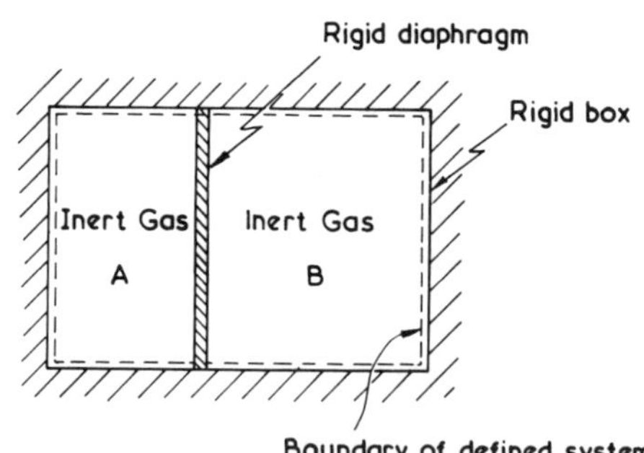

Fig. 2.3 Rigid box divided by a rigid diaphragm

If we were to relax this internal constraint by piercing a hole in the diaphragm we would widen the range of Allowed States and the condition of the system would change from a known initial Stable State to an unstable state. If thereafter isolated from its environment, the system would then settle to some final Stable State, with a uniform mixture of gases A and B on both sides of the ruptured diaphragm at a certain temperature and pressure. Moreover, on reproducing the experiment, the isolated system would again settle to this same final Stable State if no further constraints on it were relaxed (i.e. if it remained within the rigid-walled vessel).

2.4.5 A gravitational field as an external constraint on a system

In the cases of the simple and compound pendulums, we have already seen that the presence of a gravitational field exerts a constraint on a system and influences the Allowed States that the system can take up while subject to this constraint. As a further illustration, we may consider a liquid–vapour mixture contained, in a state of stable equilibrium, in a rigid box (the mixture would then be a mixture of *saturated liquid* and *saturated vapour*, as defined in Section A.3 of Appendix A to Chapter 7).

In the presence of the Earth's gravitational field, we find that, when such a mixture is macroscopically at rest in a state of stable equilibrium, the denser liquid phase occupies the lower part of the box and the less dense vapour phase the upper part. In the absence of a gravitational field, on the other hand, both the liquid and vapour would be uniformly dispersed throughout the entire volume of the box when in a state of stable equilibrium. The presence of a gravitational field thus prevents the liquid from occupying the upper part of the box when the system is in a Stable State. Thus externally imposed conservative* force fields, whether gravitational, electric, or magnetic, constitute external constraints on a system.

2.5 Definitions of Constraints, Allowed States, and Constrained Systems

The terms *Constraints, Allowed States*, and *Constrained Systems* bear particular connotations in our present usage of them. With the foregoing examples in front of us, we are now in a position formally to define these terms.

With respect to the term *Constraint*, as used here, we may note from the foregoing examples that, whether internal or external, a Constraint imposes on the system a *physical parameter* of specified finite magnitude. In the case of the simple pendulum, we had a given *radius* limiting the distance of the bob from the axis of rotation and also a gravitational field of specified *field strength*, while the air was confined within a specified *volume* by the rigid containing wall. In the example discussed in Section 2.4.4, the diaphragm divided the total volume in a fixed *ratio*. We may thus define a *Constraint* in the following terms:

Definition

A *Constraint*, whether internal or external to a system, is a restriction which, in accordance with the laws of mechanics, prevents a system from taking up states which could otherwise be accessible to it, the restriction being defined in terms of a physical parameter of specified finite magnitude.

---

* A force field is *conservative* if $\int F\,ds$ taken between any two points in the field is independent of the path taken between the points, where ds is the distance measured in the direction of the force F.

16

The term *Allowed States* may then be defined as follows:

Definition

The *Allowed States* of a system in a specified situation are all those states which are *potentially accessible* to the system (through interaction with its environment or otherwise) *while all of the Constraints to which it is subject remain unaltered.*

A prescription of all the Constraints on a system is thus a means of prescribing the totality of Allowed States that are theoretically accessible to the system while none of the Constraints are altered.

For the sake of brevity, we shall find it convenient to use the term *Constrained System* in the following special sense:

Definition

A *Constrained System* is one which is subject to specified Constraints, *including confinement within a fixed bounding surface*, and for which *none of the Constraints to which it is subject are altered during the process under consideration.*

We shall commence our study of thermodynamic equilibrium by reference to such Constrained Systems.

2.6 The nature of an adiabatic wall

As noted in Section 2.4.1, an *adiabatic wall,* as commonly understood, is one which acts as a perfect thermal insulator and so prevents heat exchange between the enclosed system and its environment. It is a hypothetical device, since we find that no real wall achieves this in practice.

Although an adiabatic wall could prevent a Constrained System from taking up certain states which would otherwise be accessible to it,* such a hypothetical wall does not constitute a Constraint within the terms of the foregoing definition. While a fixed bounding surface imposes on a system a fixed, finite physical parameter, namely a specified volume, there is no corresponding finite physical parameter associated with a perfectly insulating wall. The latter is merely a hypothetical, idealized abstraction to which a real wall approaches when it is virtually (though never, in practice, completely) non-conducting. An adiabatic wall would be of infinite, not finite, thermal resistivity; it thus represents only a theoretical concept which we find convenient to postulate for the purpose of denoting a system boundary across which no heat could flow. Internal and external Constraints, in the sense here defined, are thus not of a thermal nature, but act as constraints by virtue of the laws of mechanics. We have, of course, yet to define *heat* in formal terms; this will be done in Chapter 6.

---

* For example, although, for a gas confined within a rigid, adiabatic wall in a Stable State, one could raise its temperature (and pressure) by the input of paddle-wheel or stirring work, one could not lower its temperature, since one could not extract paddle-wheel work from it.

2.7 The nature of thermodynamic equilibrium

Although, in the foregoing discussion, we have given examples of Stable States, namely states of stable equilibrium, we have yet to define such a state. Before doing so, however, we first turn to a discussion of the nature of thermodynamic equilibrium and then give examples of states in which a system appears to be in a condition of stable equilibrium but is, in fact, in a *metastable* state.

Discussion of the different kinds of equilibrium of thermodynamic systems is frequently introduced by comparing them with the different kinds of *mechanical* equilibrium exhibited by a sphere resting on surfaces of various shapes (concave, flat, convex, and doubly curved). This, however, draws attention away from the essentially dynamic nature of *thermodynamic* equilibrium.

If we consider the chemical reaction equation

$$2\,CO + O_2 = 2\,CO_2, \tag{2.1}$$

it is, in fact, more informatively written as

$$2\,CO + O_2 \rightleftharpoons 2\,CO_2. \tag{2.2}$$

This is because, at any given temperature and pressure, there is a continuing, dynamic two-way interaction between carbon monoxide and oxygen to form carbon dioxide, on the one hand, and carbon dioxide breaking up (*dissociating*) to form carbon monoxide and oxygen, on the other. A state of thermodynamic equilibrium is reached when the rate at which carbon monoxide and oxygen are combining to form carbon dioxide is exactly counterbalanced by the rate at which carbon dioxide is decomposing. When that point is reached, all three chemical species will be present in the equilibrium mixture, even though the proportion of any one of them in this equilibrium mixture may be quite minute (e.g. at atmospheric temperature and pressure, there would be an exceedingly small amount of carbon dioxide present if one started with a mixture of carbon monoxide and oxygen). Particularly at high temperatures, the balance point of this two-way reaction shifts significantly with change in temperature and, to a much smaller extent, with change in pressure. Thus the state of stable equilibrium at a given temperature and pressure is here a dynamic state of equilibrium. The phenomenon of *dissociation* will be studied in some detail in Chapter 19.

The thermodynamic state of stable equilibrium discussed in the foregoing example is a true state of equilibrium. There are, however, other conditions which commonly come under the name of equilibrium states in spite of the fact that the system is not in a true state of stable equilibrium; such are the states of *metastable equilibrium* of which we are about to give a few examples. We describe these briefly as *Metastable States.**

---

* Some would prefer to use the term *quasistable states* for the states discussed here.

2.8 Metastable States

As a first example of a Metastable State, we may consider a spatially homogeneous mixture of two chemically reactive gases. In particular, we will consider a mixture of two parts by volume of hydrogen and one part of oxygen, contained in a vessel at ambient temperature and pressure. It is well known that, while isolated from its environment, this mixture will remain virtually unaltered in chemical composition for an almost indefinite period. However, one only needs the minute external stimulus of an electric spark to bring about ignition and the consequent transition of virtually all the hydrogen and oxygen to water substance, according to the reaction equation

$$2\,H_2 + O_2 = 2\,H_2O. \tag{2.3}$$

It is thus apparent that, although the system appeared to be in a state of stable equilibrium prior to the passage of the spark, it was in fact held by some internal constraining influence, or *passive resistance*,* in a Metastable State, from which it was triggered by what we may term the *catalytic action* of the minute stimulus provided by the spark. A surface of platinum black can also serve as a catalyst for the initiation of some chemical reactions from initial Metastable States, the catalyst itself remaining unchanged when the reaction is over.

In relation to the foregoing example, it is necessary to point out that, although it appears that the state of the hydrogen–oxygen mixture would remain unchanged for an indefinite period of time if no spark were provided, this is only because we are not prepared or able to wait long enough. As Pippard[5] has pointed out, for virtually complete transformation to water substance without the intervention of a spark, 'the time involved may be so enormous, perhaps 10^{100} years or more, that the process is not perceptible'. However, we must take cognisance of the fact that the hydrogen–oxygen mixture was not in a true Stable State; moreover, we shall have to beware of assuming that systems are in Stable States when they are, in fact, suspended in Metastable States of so-called 'equilibrium'. We must also note that the constraining influence, which we termed a 'passive resistance' and which appears to hold the system in its Metastable State, is *not* a Constraint within the terms of the definition given in Section 2.5, because it does not *prevent* ultimate transition to a Stable State. Its existence is, in fact, associated with the exceedingly slow natural *reaction rate* at the prevailing temperature and pressure.

Two further examples of a system residing in a Metastable State through the existence of some internal 'passive resistance' are provided respectively by a *supersaturated vapour* and a *superheated liquid*. If steam which is theoretically on the point of condensing (namely at the *saturation temperature* corresponding to the pressure, as discussed in Appendix A to Chapter 7) is expanded very rapidly to a

---

* Some authors apply the term *passive resistance* to what we call an internal Constraint, but we here apply the term only to the constraining influence associated with a Metastable State. If we choose to ignore the exceedingly slow reaction rate and so treat this passive resistance as a true Constraint *preventing* the reaction from proceeding, we shall then be able to treat the system as being in a Stable State, though we must always bear in mind that this is not strictly the case.

lower pressure, so reducing its temperature, the initiation of condensation may not occur immediately the pressure starts to fall, as one might expect; instead, condensation may not start until a noticeably lower pressure is reached. Over the intervening period, the vapour is then in a metastable *supersaturated* state whose departure from a state of stable equilibrium increases as the pressure drops. The metastability will ultimately be destroyed when this departure passes a certain level; in the interregnum, the metastability could also be destroyed if sufficient condensation nuclei (e.g. dust or electrically charged particles) were introduced to provide the necessary 'catalytic' stimulus. Similarly, low-temperature liquid hydrogen whose pressure is suddenly lowered in a bubble chamber remains as a *superheated* liquid, at a temperature in excess of the boiling point at the prevailing pressure, until passage of a charged particle through it provides the nucleus for local vapour formation, leading to the appearance of a visible bubble around the particle. The superheated liquid is consequently in a Metastable State.

Glass provides yet another example of a substance existing in a Metastable State. Although to all appearances glass is a solid, it is in fact a *supercooled liquid*. That it is in a Metastable State is evident from the sudden crazing of an automobile windscreen over its entire area simply in consequence of the 'catalytic' stimulus provided by the impact of a very small stone on a minute fraction of the total area of the windscreen.

As a final example of a Metastable State, we may consider a mixture of different reactive chemical species coexisting in what is commonly described as a state of *frozen equilibrium*. Taking as an example the reaction represented by equation (2.2), we have already noted that carbon monoxide, oxygen, and carbon dioxide are all present together in a stable equilibrium mixture at a given temperature and pressure and that the equilibrium composition varies with temperature. Thus, if we start with 1 mole of carbon dioxide at atmospheric temperature and pressure and heat it at constant pressure to a temperature of 3000 K, we shall find* that some of the carbon dioxide will dissociate. We shall also find that the stable equilibrium mixture at this temperature will contain about 0.56 mole of carbon dioxide, 0.44 mole of carbon monoxide, and 0.22 mole of oxygen. If we then cool the mixture slowly to the original temperature, virtually all the carbon monoxide and oxygen will recombine to form carbon dioxide. On the other hand, if the equilibrium mixture at 3000 K were cooled very rapidly, there would not be time for any significant recombination of carbon monoxide and oxygen to take place before the temperature again reached atmospheric. At that temperature the reaction rate is so exceedingly small that the mixture would remain virtually 'frozen' at a composition corresponding closely to the stable equilibrium composition at 3000 K. By contrast, in the stable equilibrium composition at atmospheric temperature there would be only exceedingly minute traces of carbon monoxide and oxygen. The condition of the rapidly cooled mixture would therefore be a Metastable State of 'frozen' equilibrium; this condition is very similar to that of the hydrogen–oxygen mixture before the passage of the spark.

* See Fig. 19.6.

Although we have discussed Metastable States at some length, our interest in such states in the present context arises only from the fact of their existence and in a recognition of the possibility that a system may be held in a Metastable State when, perhaps, we had unsuspectingly thought it to be in a Stable State. It is these Stable States that are the prime concern of equilibrium classical thermodynamics. Consequently, we have no need to attempt a formal definition of a Metastable State, the nature of which will be sufficiently evident from the foregoing discussion of some specific examples. We do, however, need a formal definition of a Stable State.

2.9 Thermodynamic states of stable equilibrium – Stable States

We are now in a position to consider in more detail the nature of that most important of all thermodynamic states, the *macroscopic state of stable equilibrium*, which, for brevity, we call simply a *Stable State*. We define it formally as follows:

Definition

A system is said to be in a *Stable State* when no finite change of state can occur unless there is an interaction between the system and its environment which leaves a finite alteration in the state of the environment.

The following important consequence follows from this definition:

Corollary

Once an *isolated system* (Section 2.3) has settled to a Stable State, there can be no finite departure from that state while the system remains isolated.

It is implicit in the foregoing definition of a Stable State that, as well as the system being in equilibrium with its immediate environment, the individual *macroscopic* parts into which the system may conceptually be divided are in mutual equilibrium with each other. If this were not the case, its state when isolated would not remain unchanged over a period of time, so that its condition would then not conform to the definition of a Stable State. Thus, when a system is in a Stable State, there is a macroscopic uniformity within the system with respect to those influences that could bring about a change of state of adjacent macroscopic parts of the system. This implies, for example, that when a fluid Simple System (Section 1.2) reaches a Stable State, any relative macroscopic motion (e.g. eddies, as distinct from molecular motion) that may have existed between adjacent parts of the system will have ceased, so that there will then be uniformity of both pressure and temperature throughout the system.

In the case of a chemically reactive system in a Stable State, there will also be uniformity of a third *potential* for energy interchange between adjacent macroscopic parts of the system; this property of a chemically reactive substance is called its *chemical potential*. For example, as we shall find in Chapter 19, in a fluid system comprising a mixture of reactive chemical species, the chemical potential of each individual species will have the same value throughout when the system is in a Stable State (though the chemical potentials of different species will not be the same). However, this measure of macroscopic uniformity then existing in such a system does not preclude the coexistence within the system of different degrees of *chemical aggregation* in adjacent parts of it; i.e. a system which comprised coexisting solid, liquid, and gaseous *phases* of a pure substance could nevertheless be in a Stable State,* when there would be uniformity throughout the system of the three potentials, namely pressure, temperature, and chemical potential. It is then customary to say that uniformity of these potentials ensures respectively *mechanical*, *thermal*, and *chemical* equilibrium.

We may now consider further the statement in Section 2.1 that equilibrium classical thermodynamics is concerned essentially with processes between Stable States. This is because in classical thermodynamics we treat matter as a continuum, so ignoring the particulate nature of matter and the quantization of energy. Thus in classical thermodynamics we can deal only with the *macroscopic* behaviour of a system. Our studies are then inevitably concentrated on Stable States, for, in order precisely to identify the state of a system when it was not in a Stable State, one would need to list the precise condition of each *microscopic* element of the system. This we cannot do when we ignore the particulate nature of matter and the quantization of energy.

In this preliminary discussion of the nature of states of stable equilibrium, we have inevitably run ahead of ourselves in the use of terms and concepts which we have not yet formally defined. It will be our task to make up for this deficiency in succeeding chapters. We next proceed, however, to consider the need for a Law relating to the existence of Stable States.

2.10 The need for a Law of Stable Equilibrium

From our discussion of the possibility of existence of Metastable States in Section 2.8, it will be evident that we can never experience absolute assurance that a system is, in fact, in a Stable State. The reason for this lack of absolute assurance lies in the fact that the system may perhaps be held suspended in a Metastable State when, from a limited knowledge of its behaviour, we had supposed it to be in a Stable State. If it is in a Metastable State, we may then either have to wait an impossibly long time or discover the appropriate 'catalytic' stimulus before we discover that it was not, after all, in a Stable State. Since we can neither afford to wait an impossibly long time nor be sure that we have identified and applied all possible 'catalytic' stimuli, we can never be absolutely sure·that the system is not in some Metastable State when we thought it to be in a Stable State. Neither have we any positive *a priori* assurance that it could take up a Stable State if left to itself.

In spite of these unavoidable uncertainties, our experimentally gained knowledge of the physical world gives us strong grounds for believing that Stable States are not

---

* This would hold at the *triple point* (Section A.4 in Appendix A to Chapter 7).

figments of our imagination and that, by comparison, Metastable States are relatively rare. However, in the absence of absolute assurance we are in the position of having to accept, as an article of faith, that a system would in time take up a Stable State when isolated from all interaction with its environment; i.e. we must accept belief in the existence of Stable States as belief in a *Law* of nature.

The need for a Law of Stable Equilibrium is further evident when one considers the implications of the description in Section 2.9 of the internal condition of a system when it is in a Stable State. There it was noted that when, for example, there is relative macroscopic motion between different parts of a fluid system, this will have ceased when the system reaches a Stable State. There is no *prima facie* case for asserting that such relative motion will always inevitably die down when such a system is isolated. Indeed, in making that assertion we rely on the *belief* that in natural processes there is what is called *'frictional' dissipation*.* Being only the expression of a belief which is incapable of formal proof, such a statement must either be incorporated in the statement of our Law or be implicit in it. If we postulate that the attainment of a Stable State is always possible, then we imply the existence of 'frictional' dissipation (or *irreversibility*) in natural processes. It is this postulate that is, in effect, expressed in the Law of Stable Equilibrium.

We have just concluded that we need to enunciate a Law which asserts that attainment of a Stable State is always possible. However, further thought reveals that we need it to go further than that. For example, when we allow a Constrained System to settle, while isolated, from a particular initial Allowed State, we would like to have assurance that, under these conditions, it would always settle to only one and the same Stable State, no matter how many times we reproduced the experiment. Belief in such a proposition is not difficult, for we could never be sure that the results of well-planned and well-executed experiments could be relied upon to be reproducible if, on one occasion, the system settled to one Stable State while, on another occasion, it settled under precisely similar conditions to a different Stable State. We therefore incorporate the proposition in our statement of the Law of Stable Equilibrium. In due course we shall find that this one addition adds very considerably to the power of the Law in enabling us to predict the macroscopic behaviour of systems.

The Law of Stable Equilibrium (LSE for short) was first enunciated by Hatsopoulos and Keenan,[1] who were also the first to demonstrate the primacy of this Law over the previously enunciated First and Second Laws of Thermodynamics. We shall demonstrate this primacy in succeeding chapters, though not following their mode of presentation in every respect.

2.11 The Law of Stable Equilibrium (LSE)

The Law was enunciated by Hatsopoulos and Keenan in the following terms:

---

* An alternative way of expressing this is to say that all natural processes are *irreversible*, a term which bears a special connotation in thermodynamics, as we shall see in Chapter 9.

Statement

The *LSE* states that *a system having specified Allowed States and an upper bound in volume can reach from any given state one and only one Stable State and leave no net effect on its environment.*

Recalling our definitions of Allowed States and Constrained Systems in Section 2.5, we may express the Law of Stable Equilibrium in the following alternative terms:

Alternative Statement

An alternative statement of the LSE is that, from a given initial Allowed State, a *Constrained System* which is permitted to undergo only such processes as leave no net effect on its environment can reach one and only one Stable State.

This statement clearly covers, as a particular case, the settling of an isolated system to a final Stable State. Embraced in the phrase 'can reach' is the implication that, given time, the system *will* ultimately reach a Stable State in the specified circumstances, though, as we saw in Section 2.8, that time may be very long indeed in the absence of appropriate 'catalytic' action. Such 'catalytic' action would, by definition, leave no finite net effect on the environment.

We may recall that, according to our definition in Section 2.5, a Constrained System is one for which none of the Constraints to which it is subject are altered during the process under consideration and that these Constraints include confinement within a fixed bounding surface.

2.12 Testing the 'truth' of the LSE

Since the LSE is a Law which we have to take on trust, we are naturally interested in the question as to how we may test its 'truth' insofar as this is possible; being a Law, its truth is not susceptible of absolute proof.

We must first note that, *so long as we do accept the LSE as a natural Law*, then we do not question its 'truth' when we encounter circumstances which appear to contradict the Law. If the behaviour of a system apparently failed to conform to the Law, we would then conclude, for example, that one or more of the following explanations might account for that apparent failure:

(a) Though we thought that the system was isolated, it was either not completely isolated or was subject to a type of interaction of whose existence we had not been aware.

(b) We had not waited long enough for the isolated system to come to a final unchanging state of stable equilibrium.

(c) On its progress towards the final Stable State, the isolated system had become lodged in a Metastable State of which we were either unaware or for the release from which we had been unable to discover the appropriate 'catalytic' stimulus.

The LSE might be said to arise from a *belief* that, at the macroscopic level, the inanimate world behaves deterministically. However, there are no means of testing such a *metaphysical* proposition against the actual behaviour of the physical world. On the other hand, the LSE is expressed in terms which enable us to remove it from the realm of a simple article of faith by introducing the element of *falsifiability* which, as Popper[6] has pointed out, is an essential ingredient of a *scientific* proposition. We do this by showing that certain propositions can be proved as *Corollaries* of the LSE. Since we then find that none of these Corollaries run counter to our experience of the physical world, as established *by observation and experiment*, we come to believe in the 'truth' of the Law, but its absolute 'truth' still remains unproven. Nevertheless, our belief in its 'truth' is notably strengthened when we discover that from it we are able to build up the complete structure of contemporary classical thermodynamics. Our confidence in the Law is further enhanced when we also discover that statistical thermodynamics enables us to identify a Stable State as the statistically *most probable* macroscopic state.

2.13 Quasistatic (reversible) processes

When a system undergoes a process at a finite rate, this inevitably results, in practice, in some departure of the system from states of true stable equilibrium. Our theoretical calculations must nevertheless relate to Stable States, since it is only for these that we have thermodynamic data. We therefore find it convenient to postulate hypothetical *quasistatic processes* in which we imagine the system to pass uninterruptedly through *a continuous succession of quasistatic Stable States*; this implies that the process would be carried out infinitely slowly, so allowing the system to settle to a Stable State at the end of each infinitesimal step in the process. Thus, in order that a gas confined within a cylinder by a moving piston could pass through a quasistatic process, it is clear that passage of the gas through a continuous succession of quasistatic Stable States could only be achieved if the environmental pressure on the back of the piston were relaxed in such a way that it was always only an infinitesimal amount less than the internal pressure; the piston would thus need to move infinitely slowly and the expansion of the gas would then be described as *fully resisted*.

A quasistatic process is an idealized process. It is, indeed, our first meeting with a very important class of idealized processes which, in thermodynamics, are called *reversible processes*.

2.14 The LSE and the source of irreversibility and lost work

The possibility that a given initial Allowed State of a Constrained System is itself a Stable State is not excluded in the statement of the LSE given in Section 2.11. In that event, the system is already in the one and only Stable State that is accessible to it while isolated from its environment and while none of the Constraints to which it is subject are altered; furthermore, the system would not be able to depart from that state while subject to those conditions. On ·the other hand, when the initial Allowed State is a non-equilibrium state, it will take time for a system, when isolated, to settle to the one and only Stable State that is accessible to it while none of the Constraints are altered. If, during passage to the Stable State, an internal 'passive resistance' were to hold the system suspended in a Metastable State, the isolation of the system would have to be temporarily interrupted to allow application of the necessary 'catalytic' stimulus required to destroy the metastability, if the system were to settle to the final Stable State within an acceptable period of time.

We have noted above that it follows from the LSE that, so long as a Constrained System in a Stable State remains isolated, it will be unable to depart from that Stable State. Thus, from a given non-equilibrium or metastable state, settlement of the isolated system to the final Stable State is a one-way or *irreversible* process. Furthermore, we can detect that the origin of this irreversibility lies in the fact that the initial condition of the system departs from a state of stable equilibrium, for there would be no need for a settling-down process if the system were initially in a Stable State. An example of such an irreversible process is that provided by a fluid when, after being stirred by a paddle wheel and thereafter isolated from its environment, it settles down to a final Stable State.

At this very early stage in our studies, we have thus already gained an insight into one of the most important conclusions in the science of thermodynamics, namely that *departure from equilibrium is the source of irreversibility* in a system. The terms *reversible* and *irreversible* bear a special connotation in thermodynamics and the study of such processes is of the greatest possible importance to the engineer because, as we shall discover in Chapter 10, *irreversibility always results in lost opportunities for producing work or in a greater work input than is ideally needed.* Unfortunately, reversible processes exist only in 'Thermotopia', that imaginary and idyllic land of the thermodynamicist in which there are no such lost opportunities.

We have already noted in Section 2.13 that a *quasistatic process*, which is one that passes uninterruptedly through a continuous succession of quasistatic Stable States, is an idealized process that would take an infinite time to accomplish. We now see that, as we then said, it is a reversible process, because at no time is there a finite departure from stable equilibrium; hence quasistatic processes do not result in any lost opportunities for producing work, or in extra work input above the ideal. In all natural processes, however, there is an inevitable departure from equilibrium, however small, since they need to be completed in a finite time. We thus come to the further very important conclusion that all natural processes are in some degree irreversible and so result in lost work or extra work input.

From the foregoing discussion, we may infer that the greater is the departure from equilibrium during a process the more inefficient is the process likely to be from the point of view of work production or absorption. We therefore need to devise a measure of irreversibility; this we shall be able to do after we have defined in Chapter 12 the somewhat abstract thermodynamic property which we call the entropy of a system.

2.15 Classical thermodynamics as an example of
the power of abstract thought

The discussion in the foregoing section was designed to set the stage for our development of the theorems of equilibrium classical thermodynamics. For this development we have to rely on the power of *abstract* thought, aided by understandings reached through *experimental* knowledge of the physical world around us. The resulting theorems are of inestimable value to the engineer, for they provide the means of setting up criteria of excellence against which to judge the performance of those work-producing and work-absorbing devices and plants on which a technologically based society is so heavily reliant. We have already noted that only *reversible* processes give maximum efficiency of work production, while all real-life processes are in some measure, however small, irreversible. Consequently, to devise, in the form of analytical expressions, performance criteria for each of our real-life devices, we can only call upon the resources of the human intellect; experimental determination of the performance of real-life devices is of no avail to this end, since the latter are always subject to some imperfection. Thus, these analytical studies will provide an outstanding example of the power of abstract thought.

2.16 Further tasks ahead

Although, as we have just noted, we have to rely on the power of *abstract* thought to derive the analytical expression for the performance criterion appropriate to a given device, when we come to work out numerical values we shall need to use *experimentally* determined data on thermodynamic property values. Before such data are convenient for engineers to use they will have to be assembled into equations, tables, and charts. Moreover, we shall find that there are theoretical relations which interconnect the values of *all* thermodynamic properties of a substance. We shall therefore need to study not only these relations but also the manner in which large quantities of experimental data on property values are assembled into *thermodynamically consistent* equations and tables. Such equations and tables will of course relate only to Stable States.

Because all real-life processes need finite time for their execution, the intermediate states through which the system or fluid passes are never precisely ones of stable equilibrium. The engineer's calculations consequently always start by dealing with idealizations of an actual process which permit him to make use of data relating only to Stable States; he later applies a correction or efficiency factor to forecast actual behaviour. Thus equilibrium thermodynamics is basically concerned throughout with idealized processes involving states of stable equilibrium.

2.17 Summary

In this chapter we have been setting the scene for our studies in the scientific discipline which goes by the name of *equilibrium classical thermodynamics*. We have seen that we shall be engaged essentially in a study of the properties and behaviour of systems as they are carried by specified processes between specified states of stable equilibrium (*Stable States*). In general, during the course of such processes, each system will be involved in interactions with other systems in its environment; it will also be subject to certain specified *Constraints*, the nature of which we have defined. We gave the name *Constrained System* to one for which none of the Constraints to which it is subject, including confinement within a fixed bounding surface, are altered during the process under consideration; to the states which are then potentially accessible to the system, either by interaction with the environment or otherwise, we gave the name *Allowed States.*

With reference to a few simple examples of thermodynamic systems, we found that when a Constrained System is isolated from all interaction with its environment it appears always to settle to one and the same Stable State from a given initial Allowed State. We found it necessary to warn, however, that a system might be lodged temporarily in a Metastable (or quasistable) State when we had supposed that it had settled to a Stable State. Since we have no *a priori* reason for supposing that any Constrained System will, when isolated, always settle to one and the same Stable State from a given initial Allowed State, we enshrined such a proposition in the *Law of Stable Equilibrium* (LSE). This was expressed in somewhat wider terms in that it related not only to isolated systems but also to systems undergoing processes which left no net effect on the environment.

We then pointed out that all natural processes require a finite time for their execution and so inevitably involve some departure of the system from states of true stable equilibrium. Since all our thermodynamic data relate to Stable States, we therefore find it convenient to postulate hypothetical *quasistatic processes* in which we imagine the system to pass uninterruptedly through a continuous succession of quasistatic Stable States. Since they must be carried out infinitely slowly, quasistatic processes are idealized processes on which we base our theoretical calculations. Noting that these represented our first meeting with a very important class of idealized processes which we call *reversible*, we concluded our discussion with a recognition of the relationship between *irreversibility* and departures from stable equilibrium. We also noted a relationship between irreversibility and lost opportunities for producing work (or in greater work input than the ideal), a factor of great engineering importance which will be studied in later chapters.

From the Law of Stable Equilibrium we shall be able to build up the closely knit logical structure of equilibrium classical thermodynamics, with its numerous theorems. In subsequent chapters we shall depict this structure in the form of a Thermodynamic Family Tree, extending the branches of the tree as we proceed chapter by chapter. As shown in Fig. 2.4, the Law of Stable Equilibrium is at the head of the tree.

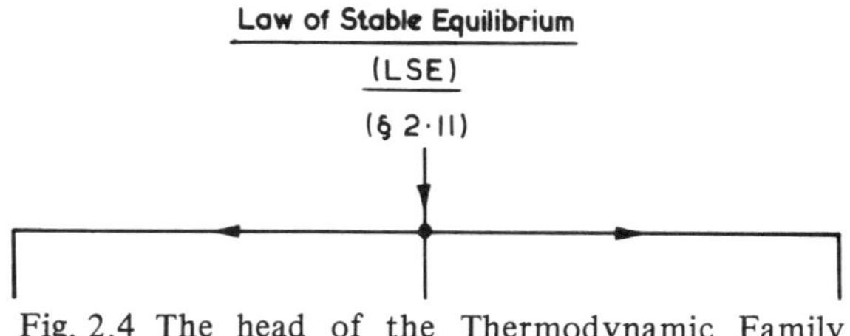

Fig. 2.4 The head of the Thermodynamic Family Tree

CHAPTER 3
Work

3.1 The concept of a pure work interaction

In Section 1.15.2 attention was drawn to the need for a definition of work which would cover the various possible modes of work transfer between systems, while in Section 1.15.4 emphasis was placed on the importance of preserving a clear distinction between the two types of interaction between systems that are known respectively as work transfer and heat transfer. We therefore need to relate our definition of work to an idealized situation in which work transfer is the only form of interaction; we shall call this a *pure work interaction.* In Chapter 6 we shall similarly consider a pure heat interaction when defining heat.

The fact that we shall define work in terms of a pure work interaction does not imply that work is not also performed in the situation depicted in Fig. 3.1, where an expanding gas clearly performs work on the moving piston while at the same time the gas is involved simultaneously in another type of interaction involving the transfer of heat to it. As was pointed out in Section 1.15.4, it is sometimes virtually impossible to separate the two types of interaction, but here at least both types are clearly and individually involved.

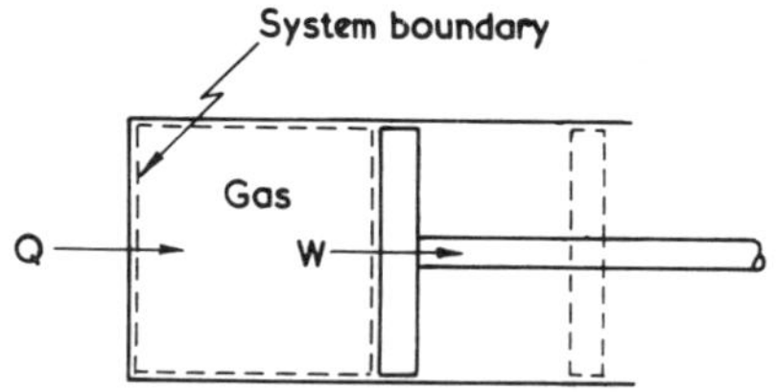

Fig. 3.1 Simultaneous transfers of work and heat

3.2 Definition of work

In the light of the foregoing comments, we define *work* as follows:

Definition

When an interaction between two systems is such that what happens in each *could* have been brought about while the *sole* end effect external *to each* was the change in level of a weight, the interaction is said to be a *pure work interaction*, in which *work* is transferred between the two systems.

It should be noted particularly that in this definition we are talking about what *could have been*; i.e. we are talking about an alternative process which would take the systems between the same states while involving as the sole end effect external to each system the change in level of a weight. Thus this is a quasistatic process (Section 2.13) in which the weight changes its level at negligible velocity. Again, this must not be taken to imply that when a weight is moved at a finite rate work is not involved; it is simply that we need to define work in terms of a situation in which it can be unambiguously identified.

It will be noted that the foregoing definition implies that it is necessary to test the effect on each system when identifying a work interaction, and not simply the effect on one of them. This necessity arises from the following circumstance. When heat is transferred from a warmer fluid to a cooler fluid, both of which are confined within fixed bounding walls and are initially in Stable States, the same warming of the cooler fluid could instead have been brought about by work input from a paddle wheel immersed in the fluid and driven by an external falling weight. However, we must not therefore infer that it was work that was supplied to the cooler fluid in the original process. That it was not work can be established by recognizing that the cooling effect on the warmer fluid could not alternatively have been brought about by an inverse paddle-wheel process involving the raising of an external weight; in Chapter 8 we shall indeed prove, as a corollary of the Law of Stable Equilibrium, that such an inverse paddle-wheel process is impossible of realization. We therefore conclude that the interaction between the two fluids in the original process could not have been a work interaction; we in fact call it a *heat interaction* and shall define this term formally in Chapter 6.

In Section 1.15.1 we recognized work transfer as a form of energy transfer, but we cannot establish this link between work and energy until we have defined *energy* in Chapter 5.

3.3 Mechanical work and the unit of work

In the study of mechanics, we express *mechanical work* as the product of a *force* exerted on a body and the *distance* through which the body is moved by that force in the direction of application of the force. Through Newton's Second Law of Motion, the force P required to give a body of mass m an instantaneous acceleration a is given by

$$P = ma. \tag{3.1}$$

In SI units, by definition of the *newton* (N), m will be in kilograms, a in metres per second squared, and P in newtons.

When a body of mass m is in a uniform gravitational field in which the gravitational acceleration is g, the mass experiences a constant downward force equal to mg. If the body is then raised infinitely slowly through a height dz by some agency, so that the sole end effect external to the agency has been the change in level of a weight, the work done by the system constituting the lifting agency will thus be given by

$$dW = mg\ dz. \tag{3.2}$$

In SI units, because the force mg is expressed in newtons, dW will be expressed in newton metres, so that the SI unit of mechanical work is the *joule* (J), since by definition of the joule, $1\ \text{J} \equiv 1\ \text{N m}$. Moreover, since we have defined work in general in terms of the change in level of a weight in a gravitational field, *the joule is the SI unit for all forms of work.*

3.4 Mechanical displacement work W_d

A further example of the performance of mechanical work is given by the work done on a piston by an expanding fluid substance confined within a cylinder by a moving piston, as depicted in Fig. 3.2(a). If the pressure p within the cylinder is greater than the environmental pressure p_0, assumed constant, then to hold the piston stationary in an equilibrium position will require the application of a resisting force P on the end of the piston rod. If we imagine the idealized situation in which this force is successively relaxed infinitesimally, the piston will move outwards infinitely slowly and the resulting expansion of the fluid will be *fully resisted*. In this ideal, fully resisted expansion, the pressure of the fluid, though falling, will remain uniform throughout the cylinder at all instants and the fluid will pass uninterruptedly through a continuous succession of quasistatic Stable States. In Section 2.13, we have already described such an idealized *quasistatic process* as being *reversible.*

The work performed on the piston *by the contained fluid* (System X in Fig. 3.2a) is described as *displacement work* W_d, and in these idealized circumstances of fully resisted, reversible expansion, we shall have

$$(dW_d)_{REV} = pA\ dx = p\ dV, \tag{3.3}$$

where p and V are respectively the instantaneous values of fluid pressure and fluid volume at the instant depicted in the figure and A is the area of the piston. Thus, this work quantity is represented by the shaded area in Fig. 3.2(b), while for a finite, fully resisted expansion between volumes V_1 and V_2 we shall have

$$\boxed{(W_d)_{REV} = \int_{V_1}^{V_2} p\ dV.} \tag{3.4}$$

As always in thermodynamics, *it is vitally important to define the system,* by delineating its boundaries, *as the first step in the analysis of the problem.* For example, if we defined System Y in Fig. 3.2(c) as our system, so that it included both the contained fluid and the piston, the displacement work performed by System Y would equal $p_0\ dV$ instead of $p\ dV$, the difference between the two being the work $P\ dx$ transmitted along the piston rod to the resisting agency.

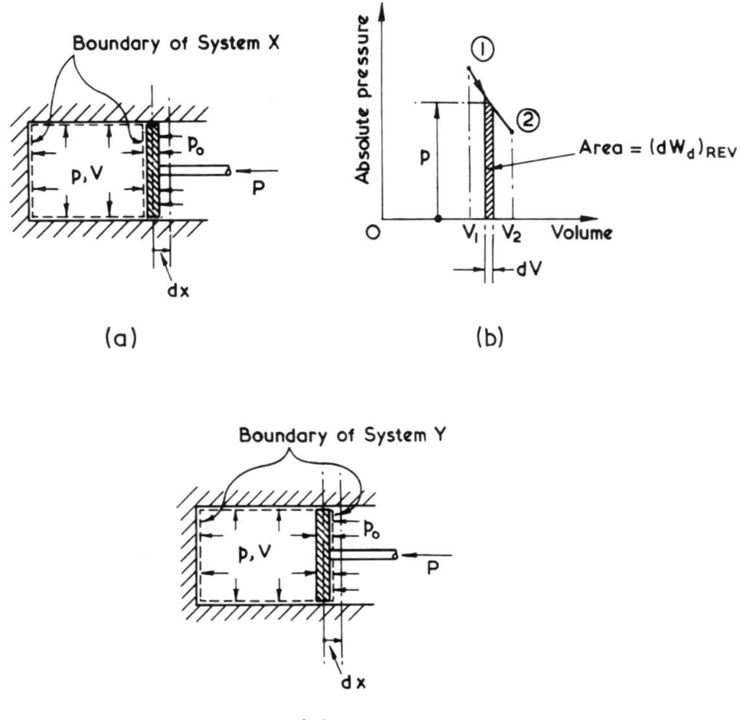

Fig. 3.2 Displacement work

It should be noted that, when the motion of the piston is not fully resisted, there will not be a strictly uniform pressure p throughout the body of the fluid, so that equation (3.4) should strictly not be used. Nevertheless, the engineer will frequently feel justified in using it; he will, for example, evaluate the *indicated* displacement work from an *indicator diagram* taken on a high-speed reciprocating engine, the *indicator* being a device which, in effect, automatically plots pressure against swept volume. For such a procedure to be permissible, the essential requirement is that the *Mach number* corresponding to the local gas velocity at any point in the cylinder should be sufficiently below unity at all times (the Mach number being defined as the ratio of the local gas velocity to the local *sonic speed*, to which reference is made in Section 18.12.4).

3.5 Other forms of work

We chose to define a work interaction in the manner set out in Section 3.2 because, as was noted in Section 1.15.2, there are other forms of work than mechanical work, notably *electrical work*. We now proceed to demonstrate that our chosen definition of work enables us to verify that the latter is indeed a mode of

32

interaction between systems which, in the present context, can be classified as a work interaction.

For this purpose, we consider in Fig. 3.3(a) an electrical storage battery supplying current to an electrical resistor and we use our definition to demonstrate that the interaction between the battery (System X) and the resistor (System Y) constitutes a transfer of work between the battery and the resistor, in spite of the fact that experience tells us that the resistor will warm up. It is true that the battery is not strictly a *system* in the sense of the definition of that term in Section 1.1, in that electrons will leave it at the negative terminal. On the other hand, under steady conditions, an equal number will return to the battery at the positive terminal in the same time; for convenience, we shall therefore regard it as permissible to use the word 'system' in this context. Hence, by showing that the flow of electrons between the two 'systems' *could* be repeated when the sole end effect external to each was the change in level of a weight, we shall in effect be demonstrating that the flow of electricity may be considered as an immaterial carrier of a work interaction between two systems.

Remembering from the discussion in Section 3.2 that the definition of a *pure work interaction* is itself related to an idealized situation, we idealize the present situation by supposing all processes to be quasistatic (so that they are carried out indefinitely slowly) and by treating the battery and leads as having negligible electrical resistance compared with that of the resistor. We shall also assume that the resistor is covered with a perfect thermal insulator and that heat conduction along the leads is negligible (using the words *thermal* and *heat* in the sense outlined in the preliminary discussion of Section 1.15.3, while deferring a formal definition of a heat interaction to Chapter 6). We consider a short time interval during which, across the boundary separating Systems X and Y in Fig. 3.3(a), System X supplies a steady electric current to System Y, causing the latter to warm up. Now, the same phenomenon would occur at the boundary if we were to make either of the following changes: (a) replace the resistor by an idealized 'lossless' motor to raise a weight, as in Fig. 3.3(b), or (b) replace the battery by an idealized 'lossless' generator

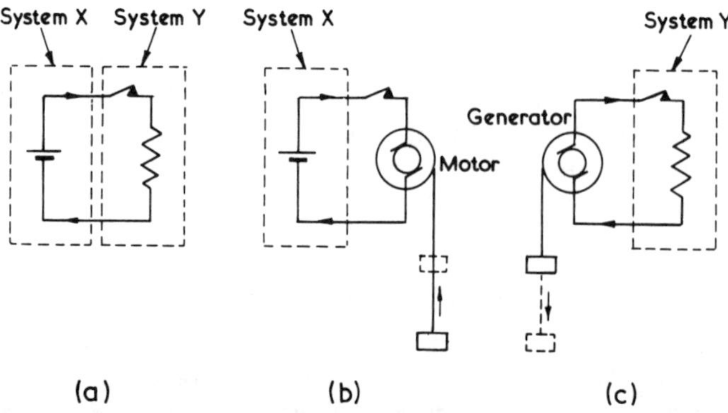

Fig. 3.3 Electrical work

driven by a falling weight, as in Fig. 3.3(c). The sole end effect external to System X and System Y respectively would in each case have been the change in level of a weight. According to our definition of work, the interaction between the battery and resistor in Fig. 3.3(a) was therefore a work interaction, and the form of work transfer in this case we call *electrical work*. We again note that work is transitory, persisting only while the interaction lasts.

3.6 Definitions of adiabatic process and adiabatic wall

The reader has already been made aware of the existence of the two modes of interaction known respectively as work and heat, though we have not yet formally defined the latter. Since we are treating only work interactions in this chapter, we need a term for processes in which work is the *only* form of interaction between systems; we call these *adiabatic** processes. Thus we have the following definition:

Definition

An *adiabatic process* is one in which the only interaction occurring between a system and its environment during the process is a work interaction.

Many authors define an adiabatic process as one in which there is no heat transfer across the boundary of a system, but the above definition is preferred because it is more specific and is perfectly precise. In any case, we have not as yet defined a heat interaction.

Having defined an adiabatic process, we find it convenient to say that, if the boundary between two systems is such that it constitutes a barrier restricting interactions between the system *solely to work interactions*, then this boundary constitutes an *adiabatic wall*. This is somewhat more precise than the frequently encountered alternative of defining an adiabatic wall as one which is impervious to the transfer of heat, a definition which was responsible for the etymological origin of the word *adiabatic*.

An adiabatic wall is a conceptual device representing a limiting case, for in the physical world it is virtually impossible completely to exclude all interactions other than work.

3.7 Summary

We may describe the science of Thermodynamics as being essentially concerned with the relationship between work, heat, and the properties of systems. In this chapter we commenced our detailed studies by defining a *pure work interaction* in

---

* It is necessary to warn that such a process has also been given the alternative title *adiathermal* (e.g. by Pippard[5]), but such usage is rare; it is also undesirable here, because we are defining such a process before we have come to the definition of a heat interaction. Pippard reserved the term *adiabatic* for what in this book, as in most texts, is described as an *isentropic* process (Section 12.4.2).

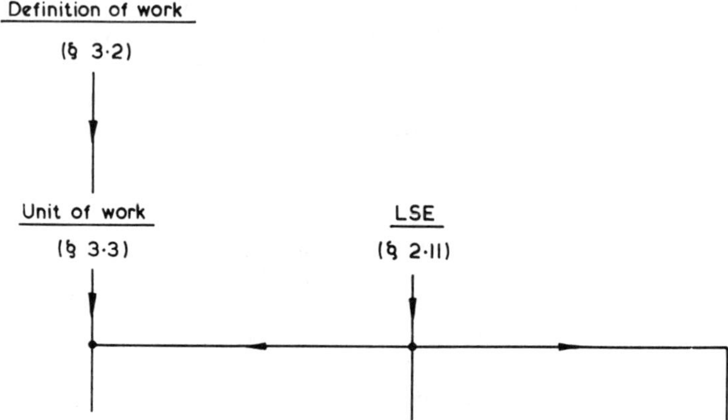

Fig 3.4 Further development of the Thermodynamic Family Tree in Chapter 3

such a way that the definition is applicable to all forms of work transfer and not just to the more familiar forms which can be recognized as mechanical work. We illustrated this generality of our definition by using it to identify the existence of another form of work, namely electrical work. These different forms share a common unit of work, the joule, which is identically equal to the newton metre. We then defined the term *adiabatic* as relating to a work-only situation, at the same time pointing out that this is a more precise use of the word than is commonly found in many other texts, which use it to imply the absence of heat transfer. We finally discussed the concept of an *adiabatic wall*.

Our Thermodynamic Family Tree now appears as shown in Fig. 3.4.

CHAPTER 4
The First 'Law' of Thermodynamics

4.1 Introduction

In the title of this chapter, the word 'Law' has been placed within inverted commas because we shall find that the conventional *Non-cyclic Statement* of this so-called Law follows as a corollary of the more fundamental proposition which in Section 2.11 was given the title of the Law of Stable Equilibrium (LSE). It therefore seems to be no longer appropriate to elevate the statement in question to the status of a Law, for that title is reserved for propositions which, though accepted as 'true', are not amenable to formal proof; we shall therefore call it Corollary 1 of the Law of Stable Equilibrium. Its importance lies in the fact that, as we shall see in the next chapter, it provides the basis for a formal definition of the thermodynamic property of a system which we call *energy*. From this Non-cyclic Statement of the First 'Law' we shall later develop the *Cyclic Statement* which is to be found in many texts; this can lay even less claim to the status of a Law.

4.2 Corollary 1 of the LSE – Adiabatic work between specified Stable States (Non-cyclic Statement of the First 'Law').

The Non-cyclic Statement of the First 'Law' relates to the work produced or absorbed by a system during an adiabatic (work-only) process between specified Stable States of the system. As a corollary of the LSE it may be expressed formally as follows:

Corollary 1 of the LSE

The work is the same for all adiabatic processes between two given Stable States of a system.

Before proceeding to the proof of this Corollary, we mention in a preliminary way an important limitation on one's freedom to bring about an adiabatic process between two specified states of a system. This limitation arises from what are commonly called Second 'Law' considerations, so that a full understanding must await our studies in a later chapter; it stems from the fact that, in all *real-life* situations, if it is possible to execute an adiabatic process which carries the system in question from Stable State 1 to Stable State 2, it will not be possible to find an adiabatic process which would carry the system back from Stable State 2 to Stable State 1. This point need not concern us further here; it will be discussed briefly in Section 4.3 and will be treated more fully in Chapter 12 during our discussion of the concept of *entropy*.

For a complete proof of Corollary 1, we need to consider two possibilities: namely case (1), adiabatic processes which involve work *input*, and case (2), adiabatic processes which involve work *output*. From what we have just said, if case (1) relates to alternative adiabatic processes between a given pair of Stable States, then case (2) must relate to alternative adiabatic processes between a different pair. In order to be perfectly general, we need to prove Corollary 1 for both types of situation; we shall prove it below for case (1) and leave case (2) as an exercise for the reader (Problem 4.2).

Considering case (1) and recalling our definition of work in Section 3.2 in terms of the change in level of a weight, we suppose that, contrary to Corollary 1, System X in Fig. 4.1 is taken from Stable State X_1 to Stable State X_2 by two alternative adiabatic processes involving different amounts of work input. Thus, in process A_{12} of Fig. 4.1(a), we suppose that the work input results from the descent of one standard weight descending through a given height from position z_1 to the stable position z_2 while interacting with System X (e.g. via paddle-work input); as a result System X will ultimately settle to some final Stable State X_2. The second standard weight there depicted meantime remains stationary. By contrast, and contrary to Corollary 1, let it be supposed that there is an alternative adiabatic process B_{12} of Fig. 4.1(b) in which two of these standard weights, instead of one, carry System X from Stable State X_1 to Stable State X_2 in descending from position z_1 to the stable position z_2.

We now let the second standard weight of Fig. 4.1(a) interact with System X (e.g. by providing paddle-work input to X) while descending from z_1 to z_2 in process A_{23}. Experimental experience* tells us that the state of System X will thereby be changed, so that it will ultimately settle to some different Stable State X_3.

Let us now define a System Y which, in both cases (a) and (b), contains within it System X and both of the weights. We then note that, from the same initial State Y_1 in both, System Y settled to Stable State Y_3 in (a) and to a different Stable State Y_2 in (b). We can conceive of the weights in State Y_1 poised at level z_1 in the manner depicted in Fig. 4.1(c), so that release from their initial positions can be achieved with no finite effects external to System Y. Thus, without any effects external to Y, we have carried System Y from the same initial state Y_1 in both (a) and (b), yet the final Stable State Y_2 in (b) is different from the final Stable State Y_3 in (a). This result violates the Law of Stable Equilibrium. We therefore conclude that the work input must be the same for all possible adiabatic processes which can take System X from Stable State 1 to Stable State 2. Corollary 1 is thus proved insofar as it relates to adiabatic processes which involve work *input*.

---

* Instead of calling upon experimental experience, we could show, as a further corollary of the Law of Stable Equilibrium, that *it is impossible to execute a process involving no other effect than the lowering of a weight*. The proof of this corollary is left as an exercise for the reader (Problem 4.1) and will be found on page 370 of reference 1.

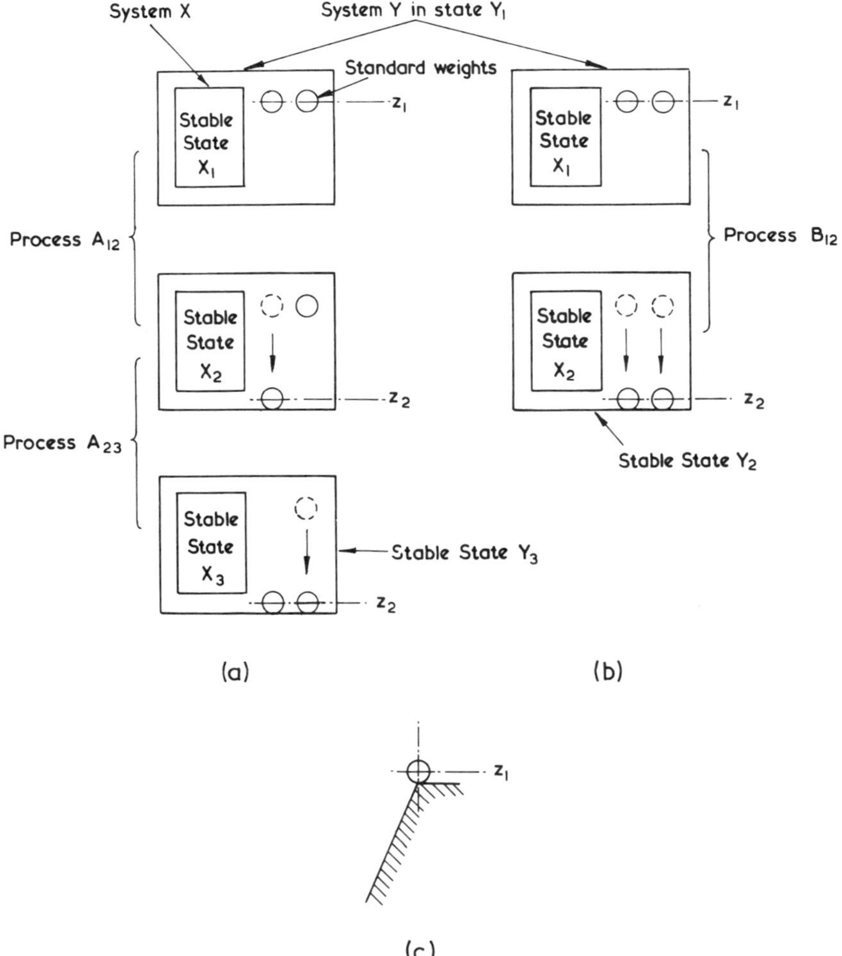

Fig. 4.1 Proof of Corollary 1 for adiabatic processes involving work input. (Based on G. N. Hatsopoulos and J. H. Keenan, *Principles of General Thermodynamics*, John Wiley & Sons, Inc., New York, 1965. Reproduced by permission of John Wiley & Sons, Inc.)

The proof of Corollary 1 is completed by demonstrating that inequality of work *output* in alternative adiabatic processes between two specified Stable States of a system would also violate the Law of Stable Equilibrium. Such a proof is left as an exercise for the reader (Problem 4.2); it was first given by Hatsopoulos and Keenan[1] and is similar to that given above, though there is an essential difference. This arises from the fact (known from experience, but proved formally in Chapter 8) that it is not possible to execute the inverse of the paddle-wheel process to produce work *output* from System X.

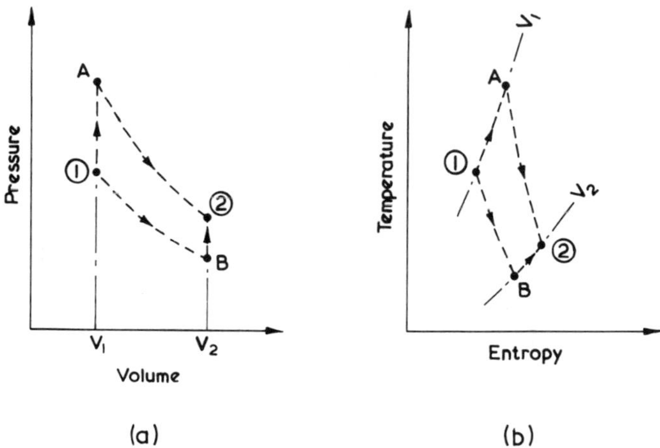

Fig. 4.2 Alternative adiabatic processes 1–A–2 and 1–B–2 from Stable State 1 to Stable State 2

4.3 An example of alternative adiabatic processes between two specified Stable States

The foregoing discussion may appear rather abstract. To add substance to it we may illustrate the significance of Corollary 1 by the following specific example involving a gaseous substance confined within a cylinder by a moveable piston. As illustrated on the pressure–volume diagram of Fig. 4.2(a), the gas could be taken from Stable State 1 to Stable State 2 by two alternative adiabatic processes, 1–A–2 and 1–B–2. In process 1–A–2, the pressure of the gas could first be raised from p_1 to p_A, at constant volume V_1, while keeping the piston fixed and putting work into the gas through the intermediary of a paddle wheel. The resisting force on the piston could then be relaxed to enable the piston to move and so allow the gas to expand to the lower pressure p_2 at the greater volume V_2, so performing displacement work on the piston. In process 1–B–2, these two operations take place in the opposite sequence, the pressure first dropping from p_1 to p_B in resisted expansion of the gas and then being raised at constant volume V_2 to pressure p_2 by means of paddle-work input. The processes are denoted in the figure by dotted lines since the precise state path between 1 and 2 in each case will not be identifiable, but each will clearly be different from the other.

It is not obvious from the figure that the net work between states 1 and 2 will be the same in process 1–B–2 as in process 1–A–2, but Corollary 1 asserts that this is necessarily so.

The following further point is worth making at this juncture, although it is somewhat premature in view of the fact that we shall not encounter the thermodynamic property which we call *entropy* until we come to Chapter 12. We shall learn in Chapter 12 that we can take a system from Stable State 1 to Stable State 2 by an *adiabatic* process only if the entropy of the system in state 2 is either greater than or equal to that in state 1, but not if it is less. Figure 4.2(b) therefore depicts the temperature–entropy diagram corresponding to the pressure–volume diagram of Fig. 4.2(a). It is evident that, although we are able to find alternative adiabatic processes taking the system from state 1 to state 2, there can be no *adiabatic* process that would take the system back to state 1 from state 2, since that would involve a decrease in entropy. This is an illustration of what we call the *irreversibility** of real-life processes.

4.4 Summary

In this chapter, using the definition of work given in Chapter 3, we have proved as our first corollary of the Law of Stable Equilibrium a proposition which is conventionally stated as one form of the so-called *First 'Law' of Thermodynamics*; this proposition states that the work quantity is the same for all adiabatic (work-only) processes between two given Stable States of a system. Before proceeding to the

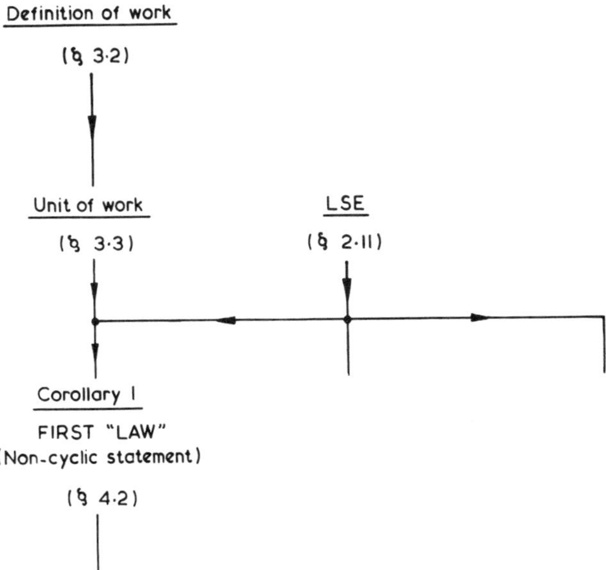

Fig. 4.3 Further development of the Thermodynamic Family Tree in Chapter 4

* The stirring or paddle-wheel process provides a particularly good example of this phenomenon, which we shall study in detail in Chapter 9. As we shall learn in Chapter 12, only in that idyllic land of the thermodynamicist, which in Section 2.14 we called 'Thermotopia' and in which all processes would be *reversible*, would an adiabatic process result in no change of entropy of a system; hence, only in an ideal process which was both adiabatic and reversible could an adiabatic process take the system back from state 2 to state 1 if an adiabatic process had taken it from state 1 to state 2.

proof of this Corollary, we pointed out that, for reasons which can only be fully understood when we have developed the concept of entropy in Chapter 12, it is not always possible to execute an adiabatic process from, say, Stable State A to Stable State B of a system; this, however, provides us with no cause for concern, for even when it is not possible to go from A to B by an adiabatic process, we shall find that it is nevertheless possible to go from B to A by such a process.

As an example of alternative adiabatic processes between two given Stable States, each involving the same amount of work, we considered the case of a gaseous substance confined within a cylinder by a moveable piston.

We shall find in the next chapter that the importance of Corollary 1 of the LSE resides in the fact that it provides the basis for a formal definition of the thermodynamic property of a system which we call *energy*.

Our Thermodynamic Family Tree now appears as shown in Fig. 4.3.

CHAPTER 5
Energy

5.1 Energy as a thermodynamic property of a system – Definition

In the last chapter we proved, as a corollary of the Law of Stable Equilibrium, that the work quantity is the same for all adiabatic processes between two given Stable States of a system. Thus, the numerical value of the work quantity will be independent of the nature of any such process and therefore dependent only on the initial and final Stable States of the system. This fact immediately suggests to us that there exists a thermodynamic property of the system whose change in value between the two specified states is related to this work quantity, for we noted in Section 1.14 that it followed from the definition of a thermodynamic property that its change in value during any process undergone by a system between specified end states is dependent only on those end states. We therefore define a new property of the system whose change in value we put numerically equal to this adiabatic work quantity. We call this property the *energy* of the system and give it the symbol E; we define it formally as follows:

Definition

Energy E is a thermodynamic property of a system whose change in value between two Stable States of the system is given by the relation

$$E_2 - E_1 \equiv - W_{12}^{a} \qquad (5.1a)$$

or

$$dE \equiv - dW^{a} \qquad (5.1b)$$

where W_{12}^{a} is the work output that would be obtained in taking the system by an adiabatic process from Stable State 1 to Stable State 2. (When an adiabatic process can only be executed in the opposite direction, W_{12}^{a} is numerically equal to the work input required to take the system from Stable State 2 to Stable State 1.)*

---

* This addendum to the definition is necessitated by the fact, to which reference was made in Section 4.3, that a real-life adiabatic process can only be carried out from a state of lower entropy to one of higher entropy. Thus, in order to raise its temperature, we can put paddle-wheel work into a fluid contained within a rigid, insulated vessel, but it is not possible for the paddle to deliver work and so take the fluid from a Stable State to some state of lower temperature.

It must be noted that one can only evaluate *changes* in energy and cannot put an absolute value to the energy of a system in a given Stable State. Consequently, in Tables of Thermodynamic Properties which list, for example, values of the *internal energy* (see Section 5.4) of a substance at different Stable States, these values represent the *difference* between the value at the given state and that at some arbitrary datum state. For example, for water substance the triple point (see Section A.4 in Appendix A to Chapter 7) is conventionally chosen as the arbitrary datum state, and the internal energy there is arbitrarily taken as being zero.

It must not be thought that, because we have defined energy in terms which relate only to adiabatic processes, we shall have no means of determining the change in energy of a system between two states unless we carry out an experiment in which we measure the work input or output in an adiabatic process between the two states. We shall see how to overcome this apparent restriction on our freedom of action after our discussion and definition of the concept of *heat* in the next chapter.

5.2 The unit of energy

From the manner in which we have defined the energy change of a system in terms of adiabatic work, it follows that the *unit* of energy will be the same as the unit of work defined in Section 3.3; in SI units this is the joule, where $1 \text{ J} \equiv 1 \text{ N m}$.

5.3 Simple Systems

By definition (Section 1.1), a thermodynamic *system* comprises a given collection of matter contained within prescribed boundaries across which no matter passes. The states which it is possible for a system to assume depend both on its own complexity and on the complexity of the influences to which it is subject. In both engineering and physics, we find it convenient initially to limit the extent of these complexities by studying first the behaviour of what we call *Simple Systems*, which were defined in Section 1.2 in the following terms:

Definition

A *Simple System* is any system which is macroscopically homogeneous and isotropic and whose *internal* condition is negligibly influenced by the effects of surface tension (capillarity), external force fields (electric, magnetic, and gravitational), and the distortion of solid phases (shear).

Since our studies will be confined very largely to Simple Systems we shall drop the qualifying adjective 'simple' when it is evident from the context that such systems are being treated.

5.4 Internal, kinetic, and potential energies of Simple Systems

The energy E of a system is the total energy of the system with respect to its energy in some arbitrary but convenient datum state. The engineer and scientist frequently break down this energy into the sum of certain precisely defined forms, such as internal energy U, gravitational potential energy, and (definitely directed) kinetic energy; together with the layman, they also make loose use of less well-defined terms, such as mechanical energy, electrical energy, chemical energy, nuclear energy, radiant energy, solar energy, tidal energy, etc. We are here concerned only with the former category of precisely defined terms, which we shall discuss in turn in relation to Simple Systems.

(a) The *internal energy* U of a Simple System is its energy in a frame of reference with respect to which it is at rest and in which it is at the arbitrary datum level (e.g. the local surface of the Earth).

(b) As we learn in the study of mechanics, when a body of mass m, starting from rest relative to the chosen frame of reference, is accelerated to a velocity V, the work that has to be performed on the body is equal to $\frac{1}{2}mV^2$. This is an adiabatic process in which the energy of the body is therefore increased by this amount. Consequently, we say that, relative to the given frame of reference, the body then has a definitely directed *kinetic energy* (KE, for short) of $\frac{1}{2}mV^2$.

In order to correct a common misconception, it is necessary to point out that, in our system of logic, when a body accelerates from rest during free fall in the Earth's gravitational field work is *not* done on the body by the Earth; the correct description of the process is to say that the kinetic energy of the body increases at the expense of its potential energy, as next defined.

(c) As we learn in the study of mechanics, when a body of mass m is raised at constant velocity through a height z in a gravitational field in which the gravitational acceleration is uniform and equal to g, the work that has to be performed on the body is equal to mgz. This is an adiabatic process in which the energy of the body is therefore increased by this amount. Consequently, we say that, if the body is at this height above the arbitrarily chosen datum level, the body then has a gravitational *potential energy* (PE, for short) of mgz.

From the foregoing definitions, it is evident that the energy E of a Simple System is related to its internal energy U through the expression:

$$E = U + \text{KE} + \text{PE} = U + \frac{1}{2}mV^2 + mgz \qquad (5.2)$$

5.5 The energy of a system in a non-equilibrium state

In Section 5.1 we defined energy in such a way that, if we were to confine ourselves only to the terms of that definition, the term *energy* would only be meaningful in relation to states of stable equilibrium, namely Stable States. We now proceed to extend the applicability of the term to cover also non-equilibrium states. To do so, we first note that energy changes of a system occur when the system suffers an interaction with its environment; in the present context, this takes the form of work output or input. We next recall that, when a system which

44

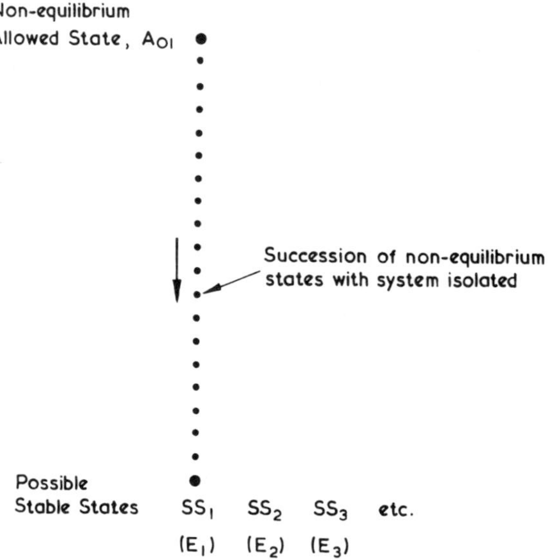

Fig. 5.1 Diagram relating to the definition of the energy of a system in a non-equilibrium state

remains subject to a fixed set of Constraints is in some non-equilibrium Allowed State (Section 2.5) and is then suddenly isolated* from its environment, the Law of Stable Equilibrium (Section 2.11) postulates that, from this particular Allowed State, the system would always ultimately settle to one and the same Stable State. In doing so, it will pass through a succession of non-equilibrium states and will leave no net effect on its environment. Thus, with reference to the diagrammatic representation of Fig. 5.1, of the many Stable States SS_1, SS_2, SS_3, etc., which it is possible for such a system to take up as a result of interactions with its environment, if on one occasion, from some given non-equilibrium Allowed State A_{01}, it were found to have settled to Stable State SS_1 after sudden isolation from its environment, it would on all other occasions in similar circumstances settle to SS_1 from A_{01}, and never to SS_2 or any other Stable State. Moreover, energy is a property and, by definition of a property, state SS_1 has associated with it only one unique value of energy, which we call E_1. Hence the non-equilibrium state A_{01}, and all the intervening non-equilibrium states (represented by dots in Fig. 5.1) through which the system passes as it settles to SS_1 during isolation, have one feature in common; from all of them, while not subject to any finite interaction with its environment, the system ultimately settles to one and the same Stable State SS_1, of energy E_1. Remembering that in adiabatic processes energy change of a system results from interaction with its environment, and noting that in the foregoing

---

* In this context, isolation does not exclude the possibility of permitting any 'catalytic' action (Section 2.8) that might be required to trigger the system from some Metastable State; such action, by definition, would leave no finite net effect on the environment.

situation there are no finite interactions, it is clearly open to us to define the energy of a system in a non-equilibrium state in the following manner, without introducing any implications as to the conservation of energy in all circumstances:

Definition

The energy of a system in a non-equilibrium state is defined as being equal to the energy of that Stable State to which the system would ultimately settle if it were suddenly isolated from its environment and the Constraints to which it was subject remained unaltered.

5.6 An alternative definition of energy

From the discussion in the preceding section, it is clear that it would be equally acceptable to define energy in the following terms:

Alternative definition

The *energy* of a system is that characteristic (i.e. property) of the system which has a common value for all those states of the system from which the same Stable State can be reached by processes which leave no finite net effect on the environment.

While this definition may appear to be more general than that given in Section 5.1, it is operationally less directly useful. To quantify the change in energy between two states of a system, we would still have to take advantage of Corollary 1 of the Law of Stable Equilibrium to equate the change in energy to the adiabatic work. This was in essence the procedure followed by Hatsopoulos and Keenan.[1] The present author prefers to define energy change directly in terms of adiabatic work, as in Section 5.1. It has already been pointed out in the last paragraph of Section 5.1 that this does not place any real restriction on our freedom of action.

5.7 Corollary 2 of the LSE – The State 'Principle'

From the discussion in Section 5.6, we may deduce what has been termed the *State 'Principle'*, but since we are able to deduce it as a logical development from the Law of Stable Equilibrium it seems inappropriate to regard it as a 'Principle' or 'Law' in its own right. Its initial enunciation by Kline and Koenig[7] predated the statement of the Law of Stable Equilibrium by Hatsopoulos and Keenan,[1] who showed that it could be derived as a corollary of the LSE. We shall find the 'Principle' directly useful in two respects, namely (a) as a justification for using energy difference as a measure of heat transfer, after we have defined the concept of a heat interaction in the next chapter, and (b) as a means of determining the number of independent variables whose value it is necessary and sufficient to specify in order to identify the state (i.e. Stable State) of a Simple System. We shall not encounter the latter use until we come to make a detailed study of the thermodynamic properties of Simple Systems in Chapter 18. The 'Principle' relates to Constrained Systems, as defined in Section 2.5, and may be enunciated in the following terms, which are somewhat more specific than those in which it was originally stated by Kline and Koenig:

Corollary 2 of the LSE – State 'Principle'

The Stable State of a Constrained System (namely one which remains subject to a specified set of **fixed** *Constraints, including confinement within a fixed bounding surface) is fully identified when its energy alone is specified.*

Expressed in another way, the 'Principle' thus declares that a system which remains subject to a specified set of fixed Constraints, including confinement within a fixed bounding surface, and which is of specified energy, can take up only one Stable State. We may verify the truth of this 'Principle' by appeal to the Law of Stable Equilibrium.

Recalling Fig. 5.1 relating to the definition of the energy of a system in a non-equilibrium state, we let Fig. 5.2 again represent diagrammatically the Stable States SS_1, SS_2, SS_3, etc. which are potentially accessible to our Constrained System through interaction with its environment while the Constraints to which it is subject remain unaltered. SS_1 is of energy E_1, SS_2 of energy E_2, etc. We then suppose that, contrary to the State 'Principle', there is another Stable State, denoted by SS_x, which is also of energy E_1. Since SS_1 and SS_x are stated to be both of energy E_1, it follows from our definition of the energy of a system in a non-equilibrium state in Section 5.5 that, while remaining isolated from its environment, our Con-

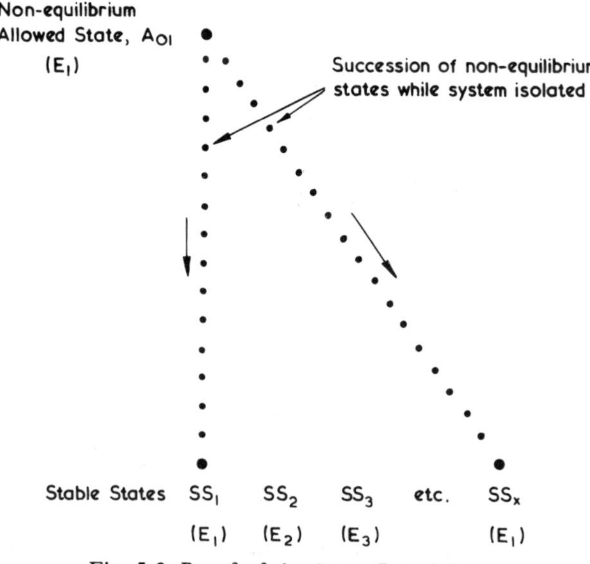

Fig. 5.2 Proof of the State 'Principle'

strained System would be able to settle to both Stable State SS_1 and Stable State SS_x from the non-equilibrium Allowed State A_{01}, of energy E_1. That, however, would violate the Law of Stable Equilibrium. We therefore conclude that in the specified circumstances it is not possible for the Constrained System to take up more than one Stable State of energy E_1. The truth of the State 'Principle' is thus verified as Corollary 2 of the LSE.

To avoid misunderstanding, we should note that it is only when the Constraints to which a System is subject remain unaltered that a given value of the energy relates to only one Stable State. For example, reference to Tables[4] of the thermo-dynamic properties of steam will reveal that the internal energy u of 1 kg of steam is equal to 2623 kJ at both the Stable State at which the pressure is 0.05 MN/m^2 and the temperature is 175 °C and at the Stable State at which the pressure is 30 MN/m^2 and the temperature is 450 °C. However, the respective Constraints to which the system is subject when in these Stable States are not the same; in the first case, the fixed bounding surface within which the kilogram of steam is confined encloses a volume of 4.12 m^3, while in the second case the fixed bounding surface encloses only 0.00674 m^3.

5.8 Summary

In Chapter 4, as Corollary 1 of the Law of Stable Equilibrium, we established the fact that the work quantity is the same for all adiabatic processes between two given Stable States of a system. Recalling from Chapter 1 that the change in value of a property of a system between specified end states is dependent only on those end states, in the present chapter we used Corollary 1 to recognize the existence of a property which we call the *energy* of a system, for which we gave a formal definition. We then identified the unit of energy as the same as that of work.

Considering the restricted case of a Simple System (namely one which is macro-scopically homogeneous and isotropic and whose internal condition is negligibly influenced by the effects of surface tension, external force fields, and the distortion of solid phases), we next related the *energy* (E) of such a system to its *internal energy* (U), its *kinetic energy* (KE), and its gravitational *potential energy* (PE) through the expression

$$E = U + KE + PE.$$

Since our definition of energy related solely to changes between Stable States of a system, we then made a logical extension of this concept to render it meaningful in relation also to non-equilibrium states. We did this by first noting that energy changes of a system occur when the system suffers an interaction with its environment; we then defined the energy of a system when in a non-equilibrium state as being equal to the energy of that Stable State to which the system would ultimately settle if it were suddenly isolated from its environment and the Constraints to which it was subject remained unaltered.

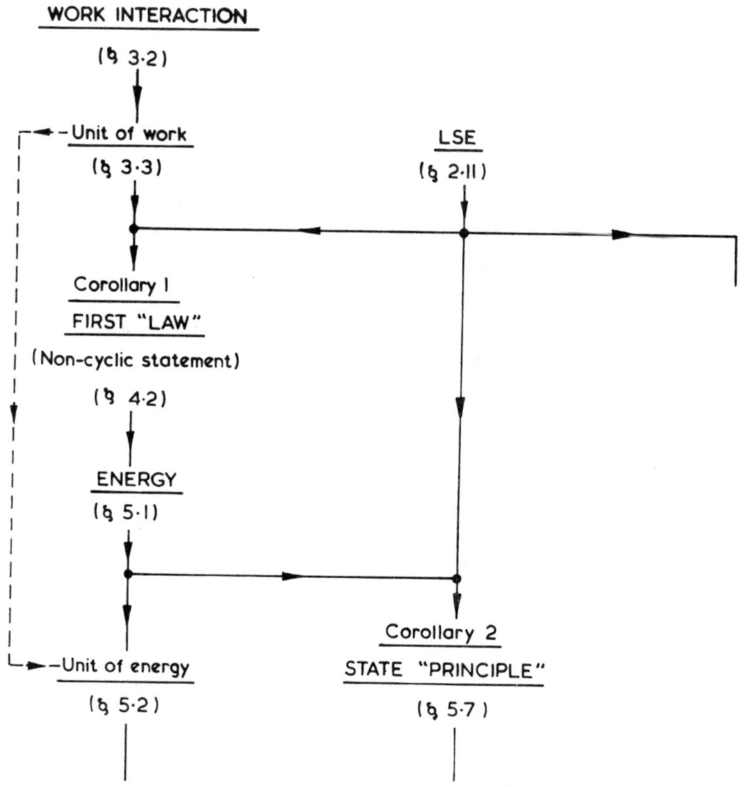

Fig. 5.3 Further development of the Thermodynamic Family Tree in Chapter 5

As a logical development from the Law of Stable Equilibrium, we finally deduced the *State 'Principle'*, which recognizes that a Constrained System* which is of specified energy can take up only one Stable State. This 'Principle' will serve us in good stead in the next chapter in justifying the use of energy difference as a measure of heat transfer, and again in Chapter 18 when we come to discuss the number of independent variables whose value it is necessary and sufficient to specify in order to identify the Stable State of a Simple System.

The adiabatic statement of the First 'Law' having provided, as Corollary 1, the first branch to stem from the head of the Thermodynamic Family Tree (namely from the Law of Stable Equilibrium), the State 'Principle' now provides a further branch and the Tree appears as depicted in Fig. 5.3.

* A *Constrained System*, by definition, is one which remains subject to a specified set of fixed Constraints, including confinement within a fixed bounding surface.

CHAPTER 6
Heat

6.1 The concept of a pure heat interaction

The necessity of making a clear distinction between the two types of interaction between systems that are known respectively as work transfer and heat transfer was stressed in Section 1.15.4. For this reason, the concept of a pure work interaction was introduced in Section 3.1. This was followed by a general definition of work in Section 3.2, while in Section 5.1 the change in energy of a system between two specified Stable States was defined in terms of adiabatic work. We thus related energy and work in such a way that one can say that work is a mode of energy transfer, noting particularly that, while energy resides in a body, work does not. In this chapter we shall define and quantify heat transfer in such a way as to identify heat also as a mode of energy transfer. To ensure an unambiguous distinction between these two modes of energy transfer, it will therefore be necessary to define a heat interaction in such a way as to exclude the possibility that the interaction in question could have been a work interaction; we shall then call it a *pure heat interaction*.

6.2 Definitions of a pure heat interaction and of heat

In Section 2.5 we gave the name *Constrained System* to one which remains subject to fixed Constraints, including confinement within a fixed bounding surface. We shall define a pure heat interaction in the following terms:

Definition

When two Constrained Systems, each initially isolated and in a Stable State, undergo a mutual interaction when brought into communication with each other while remaining otherwise isolated, that interaction between the systems is said to be a *pure heat interaction* (the energy transfer between the two systems during the interaction being described as *heat*).

The last sentence of this definition is placed in parenthesis because we have yet to establish that, when there is such an interaction between two systems, the change in energy of either system may be used as a measure of the heat quantity transferred.

We know from practical experience that when such an interaction occurs between two such systems, A and B, then one will be initially warmer than the other, as indicated by some arbitrary thermometer. We will suppose that A is

50

initially warmer than **B**. That the interaction could not have been a work inter-action may be established in the following manner.

We know from practical experience that, from its initial Stable State, the cooler System B could alternatively have been taken through the same overall change of state by means of work input arising, for example, from the lowering of a weight to rotate a paddle or friction wheel. However, we also know from practical experience that, with Constrained System A starting from an initial Stable State, the cooling effect of A could not alternatively have been brought about by the inverse of such a process. Hence the sole effect external to A could not alternatively have been the raising of a weight. Not only do we know this from experience, but we shall be able to prove it in Chapter 8* as a corollary of the Law of Stable Equilibrium. Consequently, the interaction between A and B could not have been a work inter-action. We therefore call it a pure heat interaction.

It must not be supposed that a heat interaction only occurs when the conditions set out in our definition are fully satisfied; it is simply that, if we are to give an acceptable definition of heat, we must do so in relation to a situation in which it can be unambiguously identified. There will be many instances in which both work and heat are involved in a given process, yet in which we shall be able separately to identify the work type of interaction and the heat type of interaction. There will also be other instances in which it will be less easy separately to identify them; such is the case when there is friction between surfaces in relative motion.

6.3 Quantitative measure of heat transfer

We shall now make use of the State 'Principle' enunciated in Section 5.7 to relate heat and energy, and so provide a quantitative measure of heat transfer.

Each of the systems described in the above definition of a pure heat interaction is a Constrained System, to which the State 'Principle' is therefore applicable. In proving this 'Principle' in Section 5.7 as a corollary of the Law of Stable Equilibrium, we established that the Stable State of such systems is fully identified when their energy alone is specified. In other words, when we take such a Constrained System from one Stable State to a second Stable State by means of a pure heat interaction, specification of the energy change alone is sufficient to identify the change in state of the system. We therefore choose to express heat transfer quantitatively by defining the heat quantity dQ flowing *into* a system as it passes between two closely adjacent Stable States during a pure heat interaction as equal to the resulting increase in energy dE of the system. Thus, by definition,

$$dQ \equiv dE_{\text{heat only}} \tag{6.1a}$$

or

$$Q_{12} \equiv (E_2 - E_1)_{\text{heat only}}, \tag{6.1b}$$

---

* It will be seen in Section 8.2 that we could have proved Corollary 3 at the same time as we proved Corollary 1. Thus, in principle, Corollary 3 precedes the definition of a pure heat inter-action. This is indicated in Fig. 6.3 at the end of this chapter.

where heat is arbitrarily taken as being positive for heat flow into a system and negative for heat flow out of a system. This is the opposite of the arbitrarily chosen sign convention for work. Both heat and work are transitory forms of energy transfer and, unlike energy, do not reside in a body.

It should be noted that the foregoing quantitative definition of heat followed as a logical development from the State 'Principle' and did not result from an appeal to any general Principle of the conservation of energy, a topic which will be discussed in the next chapter.

6.4 The unit of heat

Since we have defined heat quantitatively in terms of energy difference, it follows that the *unit* of heat will be the same as the unit of energy quoted in Section 5.2, and this in turn is the same as the unit of work. Thus, in SI units, the unit of heat is the joule, where $1 \text{ J} \equiv 1 \text{ N m}$.

6.5 The nature of heat transfer

We have expressed the quantity of heat transferred between two systems in terms of the change in the energy content of either system. However, the driving 'force' for heat transfer is not, of course, energy difference. We may readily verify this by considering two systems of the same material but of different size. When they are in the same state the energy content of the larger system is clearly greater than that of the smaller. On the other hand, if the larger system is slightly cooler than the smaller, we know experimentally that, on contact between the systems, heat will flow to it and increase its energy further. Thus, we know experimentally that a heat type of interaction between systems occurs when a warmer body is brought into contact with a cooler body; i.e. such an interaction takes place when the two bodies are of different 'temperature', where for present purposes the latter may be expressed arbitrarily in terms, for example, of the level of mercury in a thermometer (Section 1.15.3). We may describe this as indicating some *arbitrary temperature, θ.* A scientific definition of temperature must await the definition in Chapter 11 of the thermodynamic property of a system which we call the *thermodynamic temperature T.*

As an illustration of a pure heat interaction within the terms of the definition in Section 6.2, we may consider as our system two rigid-walled vessels and their contained fluids, the vessels being covered with material which acts as a perfect thermal insulator, so isolating each system from all interaction with its environment. As depicted in Fig. 6.1(a), we suppose that the initial Stable States of the two systems are such that a mercury thermometer in System X records a higher level θ_X than the level θ_Y recorded by an identical thermometer in System Y. If, as in Fig. 6.1(b), we bare small areas of adjacent faces of the containers and bring the two systems into communication through a small, thermally conducting rod, itself covered with a perfect thermal insulator, we shall find that ultimately the two thermometers will settle to the same level θ_Z, intermediate between θ_X and θ_Y.

52

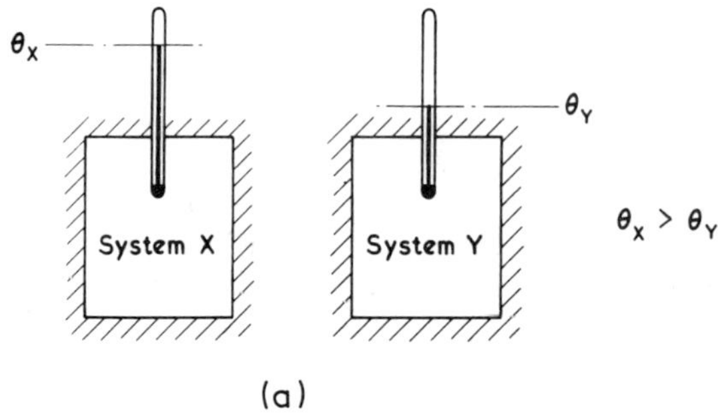

(a)

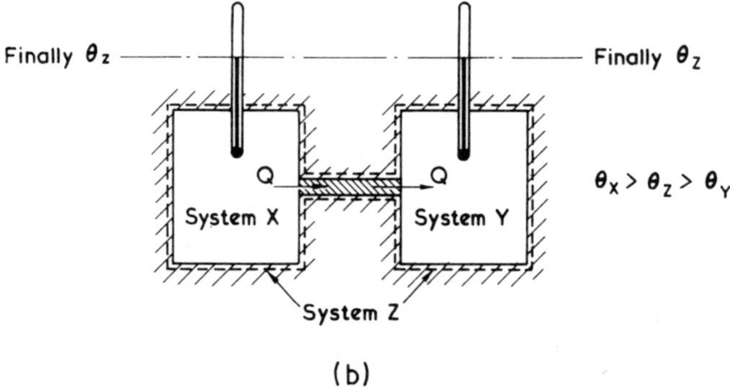

(b)

Fig. 6.1 Illustration of a heat interaction

System Z will then have settled to a final Stable State in which X, Y, and the rod are in mutual equilibrium with one another. During this process, both System X and System Y remained subject to fixed Constraints so that neither fluid was capable of work production, and there was no paddle-work input; hence the interaction between them that brought about these changes of state could not have been a work interaction. We in fact describe the process as a *heat interaction* through the transfer of *heat* along the rod.

In ascertaining the nature of the phenomenon that we have termed heat transfer, we recall that the same change of state of System Y while its arbitrary temperature rose from θ_Y to θ_Z could have been brought about by means of adiabatic paddle-work input, as depicted in Fig. 6.2; moreover, from our definition of energy in Section 5.1, we know that the energy of the system would then have increased by an amount equal to the paddle-work input. Consequently, when, in the heat interaction, we defined the heat input quantity as being equal to the increase in energy of the system, we identified it as being, *in this respect*, of the same nature as work transfer; indeed, the result of doing so was to identify the heat quantity required to bring about the given change of state as being *equivalent to* the adiabatic work quantity required to bring about the same change of state. However,

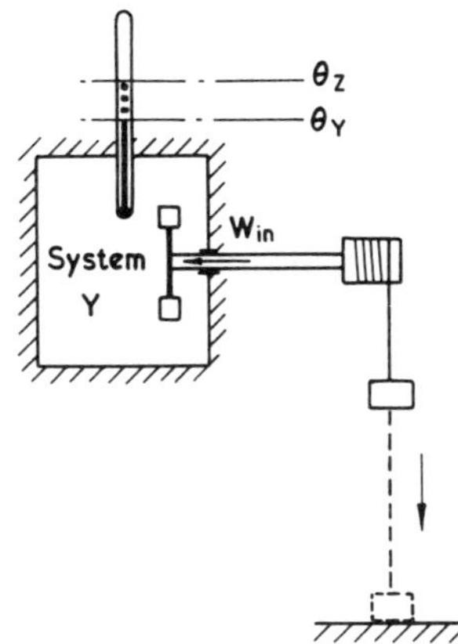

Fig. 6.2 Change of state of System Y by work transfer

we have to note the vital point that, whereas we know that System Y can be warmed by means of stirring or paddle-work input, we know from experience* that there is no comparable process by means of which System X could have been cooled while the Constraints to which it was subject remained unaltered, since we cannot then get work output from it. Hence, we again conclude that the process depicted in Fig. 6.1(b) could not have been a work interaction and that what we have called a heat interaction must be essentially different in character. The equivalence of the heat input to System Y in Fig. 6.1(b) and the adiabatic paddle-work input in Fig. 6.2 arises only from the fact that we first defined energy change of a system in terms of adiabatic work and then defined heat in terms of energy change, thus enabling us to describe both work and heat as modes of energy transfer. We justified this procedure by appeal to the State 'Principle', which we proved as a corollary of the Law of Stable Equilibrium. In none of this was there an appeal to any general Principle of the conservation of energy.

6.6 The concept of temperature difference

In the preceding section we took advantage of the reader's existing knowledge of the physical world, as gained by experiment, in order to introduce the idea of an arbitrary temperature recorded by an instrument such as a mercury-in-glass thermometer. It must be noted, however, that this was in no way an essential element of our formal definition of a heat interaction in Section 6.2. However, having defined the latter without mentioning temperature, we are now in a position to give a qualitative definition of the concept of *temperature difference* as follows:

Definition

When two finite systems which are in thermal communication with each other, but otherwise isolated, undergo a heat interaction at a finite rate, they are said to be

---

* Furthermore, we shall give a formal proof of this in Chapter 8, as Corollary 3 of the Law of Stable Equilibrium.

54

unequal in temperature. When eventually they reach a state of mutual equilibrium and the interaction ceases, they are said to be *equal in temperature.*

This is a purely qualitative definition. A scientific quantitative measure of *temperature* must await the definition in Chapter 11 of the property called *thermodynamic temperature.*

In passing, we may note that we have defined work, energy, heat, and temperature difference in that order, whereas there are other texts which follow a different sequence. For example, some texts define heat as 'that which transfers' between two systems as the result of a temperature difference, so that their sequence is work, temperature difference, heat, and energy; this is the result of following a less strictly logical form of presentation.

6.7 Heat transfer during passage between non-equilibrium states

In Section 6.2 we chose to define heat in relation to what we called a pure heat interaction, in which each system was initially in a Stable State. We did this in order to eliminate the possibility of work interaction between the systems during the process. However, it is clear that even when Systems X and Y of Fig. 6.1 are not initially in Stable States before they are brought into communication with each other, there will nevertheless be a heat type of interaction between the systems after they are brought into communication. Our definition of heat does not inhibit us from evaluating, in principle, the quantity of heat transferred in such circumstances. To do this, we have only to ascertain to what Stable State each system would ultimately have settled from its initial non-equilibrium state while remaining isolated from its environment. We then make use of the definition, given in Section 5.5, of the energy of a system in a non-equilibrium state.

A mental process such as that described above is frequently undertaken unconsciously by the engineer. When, for example, one is evaluating the heat exchanged by a fluid while passing through a heat exchanger, the fluid entering and leaving the heat exchanger will not be precisely in a Stable State. Nevertheless, one imagines the fluid to be coming from and issuing into spaces of large volume where it is nearly at rest, and in order to evaluate the property values one uses Thermodynamic Tables which quote values relating only to Stable States. It is advisable, however, always to ask oneself whether such a practice is valid in the context of the process under consideration.

6.8 The irreversibility of heat transfer

When two systems which are of finitely different temperature are brought into thermal communication with each other, but remain otherwise isolated, the combined system will for a while pass through a succession of non-equilibrium Allowed States after such communication is established. While remaining isolated, this combined system will ultimately settle to some final Stable State. From the preliminary discussion in Section 2.14 of the source of irreversibility, it will be recognized that, just as the settling down of a fluid after stirring by a paddle involves an *irreversible* process, so does heat transfer between bodies of finitely different temperature. This aspect of heat transfer is of great importance to the engineer because, as was indicated in Section 2.14, irreversibility always results in lost opportunities for producing work or in a greater work input than is ideally needed. We shall examine this more fully after we have discussed the concepts of thermodynamic irreversibility and reversibility in Chapter 9.

6.9 Summary

Just as in Chapter 3 we defined work in circumstances which enabled us to identify a pure work interaction, so in the present chapter we have defined heat in a different set of circumstances which enable us to identify a *pure heat interaction.* To do this, we had to exclude the possibility of the interaction in question being a work interaction, so that we defined a pure heat interaction as one which takes place between two Constrained Systems, each of which was initially isolated and in a Stable State before being brought into communication with the other. We then noted that the proposition known as the State 'Principle', which we developed in Section 5.7 as a corollary of the Law of Stable Equilibrium, enabled us to deduce that, when a Constrained System is taken from a given Stable State to a second Stable State by means of a pure heat interaction, specification of the energy change alone is sufficient to identify the new Stable State of the system. This fact allowed us logically to express the heat quantity flowing into a system during a pure heat interaction as being equal to the resulting increase in energy of the system. Without recourse to any so-called 'principle of the conservation of energy', this procedure enabled us to identify the unit of heat as being the same as that of energy, which we had previously identified in Chapter 5 as being the same as that of work.

Following a discussion of the nature of heat transfer, we were enabled to give a qualitative definition of the concept of *temperature difference* in relation to heat interactions, at the same time noting that a quantitative, scientific measure of *temperature* would have to await the definition in Chapter 11 of the property called *thermodynamic temperature.*

Since we had defined a pure heat interaction in relation to processes between states of stable equilibrium (Stable States), we next made the point that there is still a heat type of interaction even when the two Constrained Systems which are brought into communication with each other are not initially in Stable States, but in non-equilibrium states. We then ended with a brief discussion on the irreversibility of heat transfer when it takes place between two systems which are of finitely different temperature.

Our Thermodynamic Family Tree now appears as shown in Fig. 6.3.

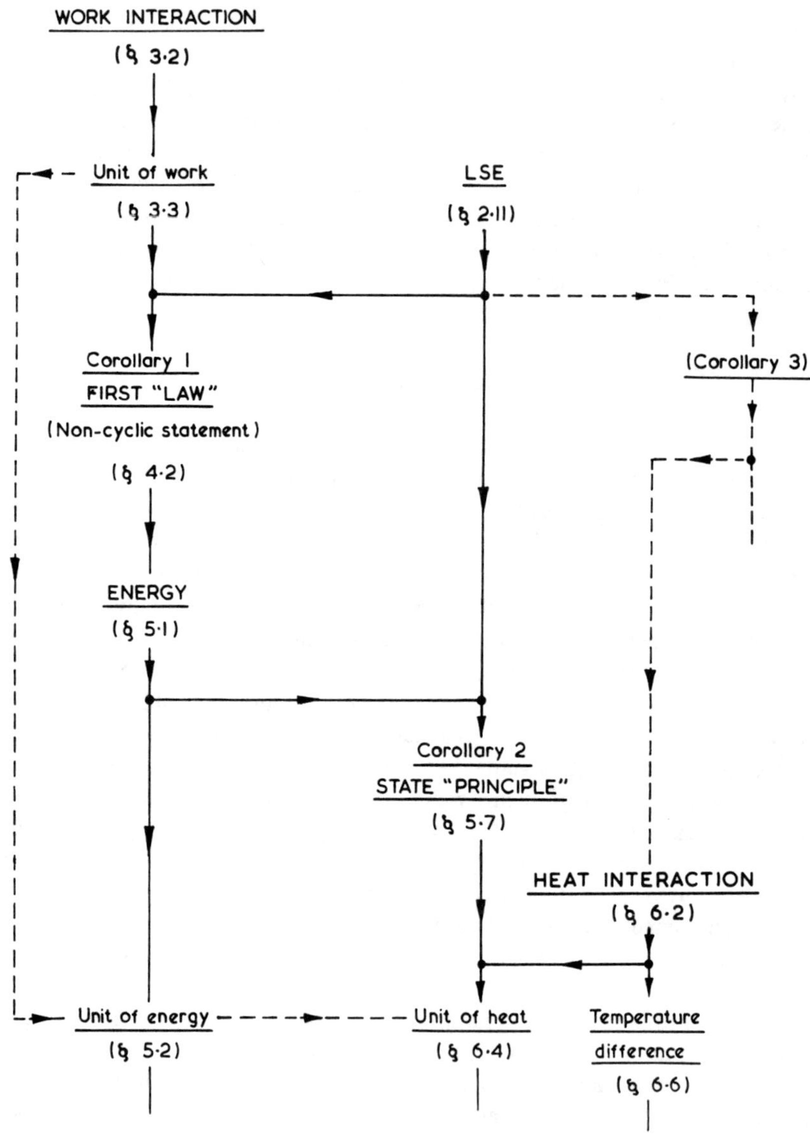

Fig. 6.3 Further development of the Thermodynamic Family Tree in Chapter 6

CHAPTER 7

Work, Heat, and Energy–Energy Conservation Equations (With Appendix A)

NON-FLOW PROCESSES

7.1 Non-cyclic processes involving both work and heat interactions

From the definition of energy in Section 5.1, we know that, when a system undergoes an infinitesimal, non-cyclic, adiabatic (work-only) process,

$$dE \equiv - dW^{\mathrm{a}}. \tag{5.1b}$$

Likewise, from our method of defining a quantitative measure for heat in Section 6.3, we know that, when a system undergoes an infinitesimal, non-cyclic, heat-only process between Stable States,

$$dQ \equiv dE_{\text{heat only}}. \tag{6.1a}$$

We will now consider System 1 in Fig. 7.1 while it undergoes an infinitesimal non-cyclic process, in the course of which it suffers both a work interaction while delivering work of magnitude dW_1 and a heat interaction while receiving heat of magnitude dQ_1; these can either occur successively or simultaneously, in which case they must be separately identifiable. We suppose that System 1 receives its heat from System 2, for which, from equation (6.1a), we may write

$$dE_2 = - dQ_1. \tag{7.1}$$

Now energy is an extensive property of a system, its change in value during any process between specified states of the system being independent of the nature of the process. Hence, from equation (5.1b), we may write for the combined System 3, comprising Systems 1 and 2 together,

$$dE_3 = dE_1 + dE_2 = -dW_1; \tag{7.2}$$

whence, from equations (7.1) and (7.2), we have the following equation relating to System 1:

$$dE_1 = dQ_1 - dW_1. \tag{7.3a}$$

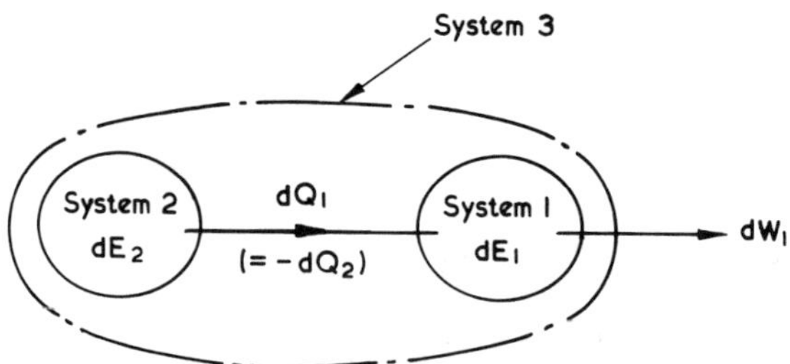

Fig. 7.1 Work and heat interactions in infinitesimal non-cyclic process

Thus, for a system undergoing an infinitesimal, non-cyclic process during which it delivers work dW to its environment and receives heat dQ from its environment, we have the equation

$$dE = dQ - dW. \tag{7.3b}$$

For a finite, non-cyclic process this becomes

$$\Delta E = \Sigma (Q - W). \tag{7.4}$$

In the special case when changes in kinetic energy and potential energy of the system are negligible during the process, we see from equation (5.2) that $\Delta E = \Delta U$, where U is the *internal energy* of the system. We may then write

$$\Delta U = \Sigma (Q - W). \tag{7.5}$$

We traditionally call equations (7.3b), (7.4), and (7.5) *Energy Conservation Equations for the system*, though it should be noted that in deriving them we have not called upon any general principle of the conservation of energy.

7.2 A note on cyclic processes in equilibrium classical thermodynamics

Before turning to a consideration of cyclic processes involving both work and heat interactions, it is worth drawing particular attention to the fact that in equilibrium classical thermodynamics it is only meaningful to treat cyclic processes which start from a Stable State.

By definition, a cyclic process must end at the same state as that from which it started. If the state at the start of a cyclic process were not a Stable State, then we would need to know the initial condition of the system in microscopic detail if we were to be able to bring the system back to the same identical condition in order to execute a completed cyclic process. However, equilibrium thermodynamics treats matter as a continuum and so is unable to say anything about the precise condition of the system at any instant in *microscopic* detail; i.e. it deals only with *macroscopic* states.

Thus, whenever we talk about a cyclic process in equilibrium thermodynamics we are dealing with a process which starts from an initial Stable State of the system in question and ends at the same Stable State. Most texts fail to note this important point.

7.3 Cyclic processes involving both work and heat interactions

Since E and U are both *properties* of a system, we note from equation (1.1) of Section 1.14 that $(\Delta E)_{\text{cycle}} = (\Delta U)_{\text{cycle}} = 0$, so that from equation (7.5) we have

$$\Sigma (Q - W) = 0, \tag{7.6a}$$

where the symbol Σ denotes algebraic summation round the cyclic process, heat rejected and work absorbed both being treated as negative. If the path can be identified throughout the cycle, we may write

$$\oint (dQ - dW) = 0, \tag{7.6b}$$

where the symbol $\oint$ denotes integration round the cyclic path.

Like the 'Energy Conservation Equations' of the preceding Section, these two equations are frequently described as expressions of the so-called *'Principle' of the Conservation of Energy*. However, we see here that they follow logically from the way in which we have defined the concepts of work, energy, and heat with the aid of the Law of Stable Equilibrium and do not rely on the postulation of any such general 'Principle'.

7.4 Cyclic Statement of the First 'Law'

Equations (7.6a) and (7.6b) have frequently been described as statements of the First Law of Thermodynamics. At the same time, we saw in Section 4.2 that the proposition there described as Corollary 1 of the Law of Stable Equilibrium has also been described as a statement of the same Law, and we therefore called it the *Non-cyclic Statement* of this so-called First 'Law'. We may then call equation (7.6a), or equation (7.6b), the *Cyclic Statement* of this 'Law', though we see that it is not a Law in its own right but a logical development of ideas stemming from the basic Law of Stable Equilibrium.

Some authors have even described equation (7.3b) as a statement of the First 'Law'. This cannot be justified, for we can see from equation (1.2) of Section 1.14 that, to those who take equation (7.6b) as the Cyclic Statement of the First 'Law', equation (7.3b) is the defining equation for the thermodynamic property E, namely energy. In our own presentation of equilibrium thermodynamics we have rejected such an approach and have preferred in Section 5.1 to give a definition of E deriving from ideas associated with the Law of Stable Equilibrium.

It is of interest to note that we derived the Non-cyclic Statement in Section 4.2 even before definition of the concepts of energy and heat. Consequently, not only might we legitimately challenge its status as a Law but we might go further by suggesting that, if it is to be called a Law at all, it might better be described as the 'First Law of Mechanics' rather than the 'First Law of Thermodynamics', since it would still be applicable in a hypothetical 'work-only' universe.

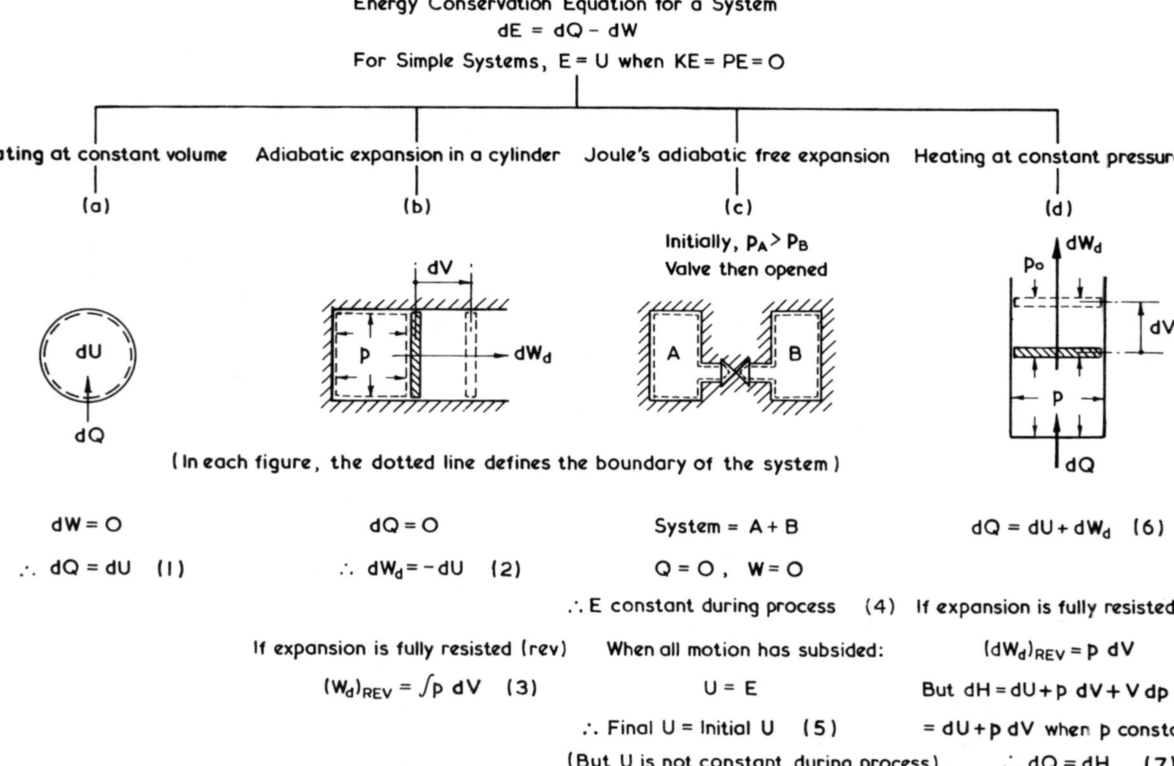

Fig. 7.2 Application of the Energy Conservation Equation for non-flow processes. (From Kempe's *Engineers Year-Book*, chapter on 'Thermodynamics' by R. W. Haywood. Reproduced by permission of Morgan-Grampian Book Publishing Co. Ltd.)

7.5 Applications of the Energy Conservation Equation for
non-flow processes — System analysis

As already noted, non-flow processes are those involving no fluid flow across the boundaries of a device or plant, which thus comprises a system. Energy calculations relating to the device or plant are thus made by application of the Energy Conservation Equation for a system (equation 7.4 or 7.5, according to circumstances). This procedure is described as *system analysis*, to distinguish it from the *control-volume analysis* which we shall apply to flow processes in the next Section. Some simple applications are illustrated in Fig. 7.2; they all relate to *Simple Systems* in the sense defined in Section 5.3, namely systems which are macroscopically homogeneous and isotropic and whose internal condition is negligibly influenced by the effects of surface tension, external force fields, and the distortion of solid phases.

It should be noted that, while case (c) in Fig. 7.2 involves flow through the valve after it has been opened, the system to which the energy conservation equation is being applied has been so defined that it contains within itself the fluids within both vessels A and B and the valve; hence there is no fluid flow across the boundary of this system. This example serves to emphasize the point that one should always carefully define the boundary of the system as the first step in the energy analysis of a non-flow process.

When shortly we come to consider flow processes, we shall need to define the boundary of a *control volume*, and for this purpose we shall use a chain-dotted line in order carefully to distinguish it from a *system* boundary, for which we here use a normal dotted line.

FLOW PROCESSES

7.6 Control-volume analysis of flow processes

The number of engineering processes which involve material flow across the boundaries of a plant (*not* a system) greatly exceeds the number not involving such flow. By definition of a system (Section 1.1), no matter may cross its boundaries, so that in dealing with flow processes we cannot directly apply the results already deduced for non-flow processes, which involve *system analysis*. Flow processes are best made the subject of *control-volume analysis*, in which we first define a *control surface* surrounding the device or plant. Both energy and matter may enter and leave the control volume contained within the prescribed control surface and before we can see, for the first time, how to treat the problem directly by control-volume analysis we must transform the control-volume problem into a system problem. We do this by considering a certain time interval and then defining a system boundary which initially surrounds both the already prescribed control volume containing the device or plant and that amount of fluid at the inlet to the control volume which will enter the latter during the prescribed time interval. By the end of the prescribed time interval, the system boundary will have moved in such a way as still to contain within it precisely the same matter as was initially within it, but it

62

will now surround both the initially prescribed control volume and any matter that has left the control volume during the prescribed time interval.

The above procedure is best understood by reference to the specific examples treated below. It leads to equations which are readily recognized as *Energy Conservation Equations for the control volume*, as distinct from the Energy Conservation Equation for a system which provided the initial starting point for the analysis. Having once realized this, we do not need always to go back to system analysis when dealing with a flow process; we can instead immediately write down the Energy Conservation Equation for the control volume. In this way, control-volume analysis is applied to a flow problem at the very outset.

7.7 Simplifying assumption regarding fluid states

In order to make the processes amenable to simple analysis in this introductory treatment, it will be assumed in all the following cases that the states of the fluid at entry to and exit from the control volumes in question are Stable States* which do not vary with time during the period to which the analysis relates.

7.8 Enthalpy – a thermodynamic property

We shall find that, since the foregoing simplifying assumption implies constancy of fluid pressure at entry to and/or exit from the control volume in steady-flow processes, a certain combination of fluid properties, namely $U + pV$, persistently recurs. This combination is clearly itself a property and, simply for convenience, we give it a special name and associated symbol, viz. the *enthalpy H*, defined as follows:

Definition

The *enthalpy H* of a system of internal energy U and volume V, at a uniform pressure p, is defined by the expression

$$H \equiv U + pV. \tag{7.7}$$

The unit of enthalpy is clearly the same as that of energy, namely the joule ($\equiv 1$ N m). It is wise not to attempt to give any more physical significance to this thermodynamic property, enthalpy, than is contained in the above equation by which it is defined.

7.9 Subclassification of flow processes

For purposes of analysis, we may conveniently classify flow processes in the present context into three separate types: namely (a) *semi-flow*, (b) *steady-flow*, and (c) *general-flow* processes. We shall also find it convenient to treat them in that

---

* More precisely described as quasistatic Stable States.

order, although in engineering plant the second type will be encountered much more frequently than the other two.

7.10 Semi-flow processes

We here give the name *semi-flow* to the type of problem in which, for example, a 'gas' bottle is recharged by connecting it to a large source of fluid at constant pressure (the word 'gas' has been placed in inverted commas because the bottle may well contain a two-phase mixture of liquid and vapour, say of carbon dioxide or nitrous oxide). We take this as our first example of a flow process because it enables us to see, with as little complication as possible, that the energy brought into the bottle by the incoming fluid under these conditions, if its kinetic energy is negligible and potential energy changes can be ignored, is its enthalpy H, and not just its internal energy U. We show this in the following manner.

With reference to Fig. 7.3(a), we first define a fixed control volume by drawing the control surface shown; this control volume virtually coincides with the bottle volume, but extends up to the inlet face of the shut-off valve. For the sake of simplicity, it will be assumed that the valve is opened in such a way that the kinetic energy of the gas passing this inlet face is at all times negligible; we shall also assume that the pressure and other fluid properties there remain constant and equal to those in the large source from which the fluid comes.

Since we have so far only an Energy Conservation Equation for a system, we next delineate the boundary of a system which initially, before gas flow starts, comprises both the contents of the control volume and the amount of fluid which will ultimately flow into it, as depicted in Fig. 7.3(b). In this situation the internal energy *of the system* will be the sum of U_{c_i} and U_1 where

U_{c_i} = internal energy within the control volume initially

and

U_1 = internal energy of the fluid about to enter the control volume.

As the recharging process proceeds, the boundary of the system whose initial position is indicated in Fig. 7.3(b) will move to the right until, on completion of the process, it will coincide with the control surface depicted in Fig. 7.3(a). This final condition is represented in Fig. 7.3(c) and then the internal energy *of the system* will coincide with that of the control volume, which we will call U_{c_f} (namely, the internal energy of the final contents of the bottle).

To the system depicted in its initial and final states in Fig. 7.3(b) and (c) respectively, we may now apply the *Energy Conservation Equation for a system* in these circumstances, namely

$$\Delta U = \Sigma (Q - W). \qquad (7.5)$$

For this system we have

$$\Delta U = (U_{\text{final}})_{\text{system}} - (U_{\text{initial}})_{\text{system}}$$

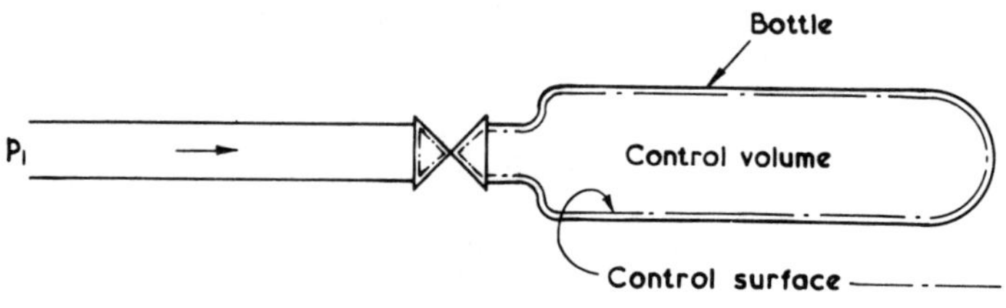

(a) Bottle re-charging process (semi-flow)

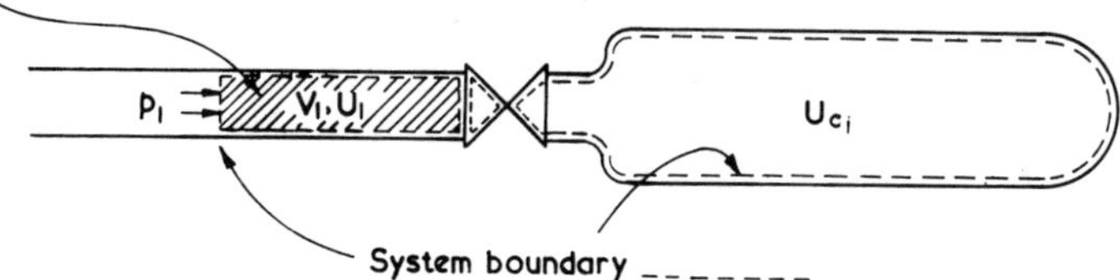

(b) System boundary initially

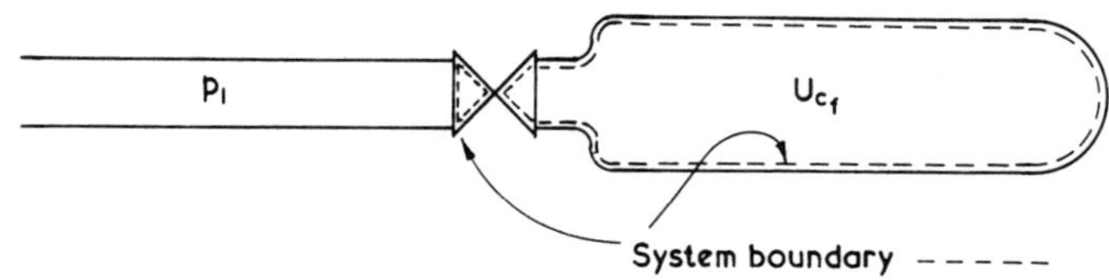

(c) System boundary finally

Fig. 7.3 Steps in the analysis of a semi-flow process

Therefore,

Now,

$$\Delta U = U_{c_f} - (U_{c_i} + U_1) \tag{7.8}$$

$\Sigma Q = 0$, assuming that the process is adiabatic,

and

$\Sigma W = -$ displacement work done *on the system.*

Therefore,

$$\Sigma W = -p_1 V_1. \tag{7.9}$$

From equations (7.5), (7.8), and (7.9) we have

$$U_{c_f} - (U_{c_i} + U_1) = 0 - (-p_1 V_1)$$

or

$$U_{c_i} + (U_1 + p_1 V_1) = U_{c_f},$$

giving, finally,

$$\boxed{U_{c_i} + H_1 = U_{c_f}\,,} \tag{7.10}$$

where $H_1 = (U_1 + p_1 V_1)$, the enthalpy of the new charge of fluid before admission to the bottle.

Equation (7.10) was derived by application of the Energy Conservation Equation for a system to the conceptual system devised for the purpose in Fig. 7.3(b) and (c). However, this is a laborious procedure, and when, *with the initial control volume in mind*, we take a look at equation (7.10) term by term, we see that it tells us that the initial internal energy within the control volume plus the *enthalpy* that came in with the fresh charge is equal to the final internal energy within the control volume.

Thus equation (7.10) is the *Energy Conservation Equation for the control volume*. From it we learn the very important fact that, *in these conditions of flow from a constant-pressure source* into a control volume, the energy brought into the control volume (in this case, the bottle) by the incoming fluid is the *enthalpy H_1*, and *not* simply the internal energy U_1. Once having recognized this important fact, we can write down Energy Conservation Equations for control volumes in other flow processes with as much ease as we write down the Energy Conservation Equation for a system. We shall follow this more direct line of attack in our treatment of the other types of flow process listed in Section 7.9.

Before proceeding to these other types of flow process, we may briefly consider the situation when it is necessary to take account of the kinetic energy (KE) of the incoming fluid, and also perhaps of the potential energy (PE) with respect to an arbitrary datum level in a gravitational field. From equation (5.2) showing the relation between E and U, we can readily see that the energy brought into the control volume by the incoming fluid would then be given by:

$$\text{Incoming energy} = U_1 + p_1 V_1 + \text{KE}_1 + \text{PE}_1$$

$$= H_1 + \text{KE}_1 + \text{PE}_1. \tag{7.11}$$

Thus, if a mass m is taken into a control volume from a constant-pressure source over a certain period during which the fluid velocity is constant at V, we shall have

$$\text{Incoming energy} = m\left(h + \frac{V^2}{2} + gz\right). \tag{7.12}$$

7.11 Steady-flow processes

For present purposes, a *steady-flow process* is defined as one in which the states and velocities of the fluid entering and leaving a rigid, stationary control volume are invariable with time and in which there is no change in the mass and

66

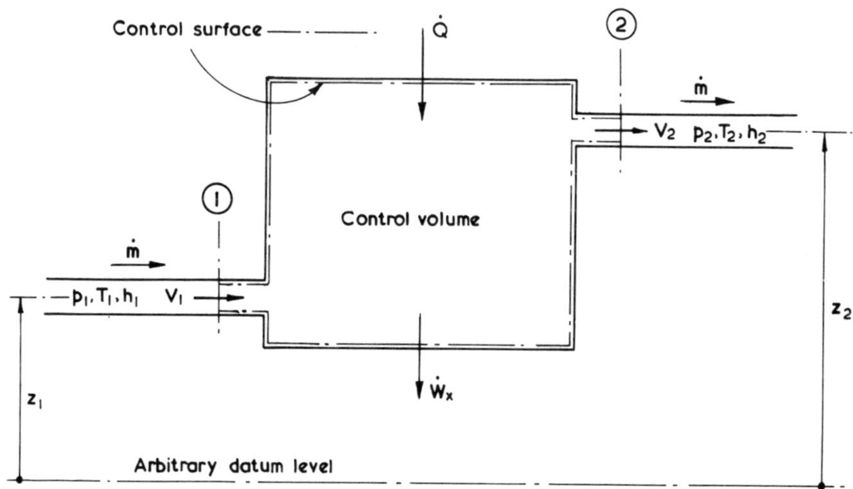

Fig. 7.4 General representation of a steady-flow process

energy within the control volume during the time interval considered. Thus the mass flow rate $(\dot{m})$ of the fluid is constant and is the same at entry to and exit from the control volume. A general representation of such a process is given in Fig. 7.4 for a device which is conceived as delivering shaft work (a term which covers electrical work, as in a fuel cell, as well as mechanical shaft work) at a constant rate $\dot{W}_x$ and taking in heat at a constant rate $\dot{Q}$. The chain-dotted line delineates the control surface which defines the control volume to which we have to apply the Energy Conservation Equation.

From the knowledge gained in our study of the semi-flow process in the previous Section, the energy brought into the control volume by the entering fluid will be the sum of the enthalpy, kinetic energy, and potential energy (equations 7.11 and 7.12). Since the device also discharges the fluid to a region of constant pressure p_2, the energy carried out of the control volume by the leaving fluid will also be the sum of these terms. Thus we shall have for the control volume:

$$\text{Rate of energy inflow} = \dot{m}\left(h_1 + \frac{V_1^2}{2} + gz_1\right) + \dot{Q} \qquad (7.13)$$

and

$$\text{Rate of energy outflow} = \dot{m}\left(h_2 + \frac{V_2^2}{2} + gz_2\right) + \dot{W}_x. \qquad (7.14)$$

From the foregoing definition of a steady-flow process, the energy within the control volume remains constant, so that the rate of energy outflow must equal the rate of energy inflow. Hence, from equations (7.13) and (7.14) we have

$$\dot{m}\left(h_1 + \frac{V_1^2}{2} + gz_1\right) + \dot{Q} = \dot{m}\left(h_2 + \frac{V_2^2}{2} + gz_2\right) + \dot{W}_x. \qquad (7.15)$$

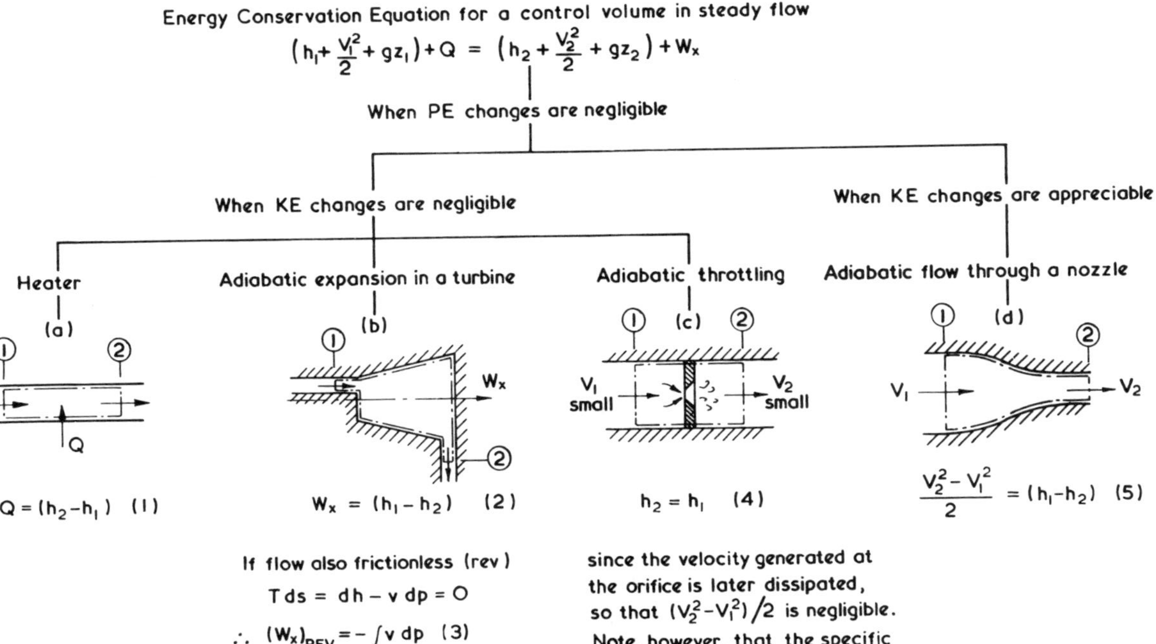

Fig. 7.5 Simple Applications of the Steady-Flow Energy Equation. (*Note:* In each figure, the chain-dotted line denotes the control surface which defines the control volume.) (From Kempe's *Engineer's Year-Book,* chapter on 'Thermodynamics' by R. W. Haywood. Reproduced by permission of Morgan-Grampian Book Publishing Co. Ltd.)

68

This *Energy Conservation Equation for the control volume* is frequently described as the *Steady-Flow Energy Equation*. By dividing through by $\dot{m}$, we may alternatively express all terms per unit mass of fluid flowing through; viz.

$$\left(h_1 + \frac{V_1^2}{2} + gz_1\right) + Q = \left(h_2 + \frac{V_2^2}{2} + gz_2\right) + W_x, \qquad (7.16)$$

where $Q = \dfrac{\dot{Q}}{\dot{m}} =$ heat supply per unit mass of fluid

and

$$W_x = \frac{\dot{W}_x}{\dot{m}} = \text{shaft work per unit mass of fluid.}$$

Some specific applications of equation (7.16) to simple engineering processes are illustrated in Fig. 7.5.

With respect to process (b) in Fig. 7.5, attention is drawn to the fact that understanding of the footnote relating to frictionless (reversible) flow must await the further development of the concepts of irreversibility and reversibility in Chapter 9 and the development of the concept of entropy (S) in Chapter 12. In the meantime, particular notice should be taken of the difference between the expression for reversible *displacement* work (in *non-flow* processes) per unit mass of fluid and that for reversible *shaft* work (in *steady-flow* processes); these are respectively

and

$$\boxed{(W_d)_{\text{REV}} = + \int p \, dv} \quad \text{(see Section 3.4)}$$

$$\boxed{(W_x)_{\text{REV}} = - \int v \, dp} \quad \text{(see Section 12.6)}$$

With respect to process (c), namely that of *adiabatic throttling*, it should be noted that as the fluid passes through the throttling restriction its velocity will increase and its specific enthalpy will therefore fall. However, on emergence from the throttling restriction into the downstream passage of much larger cross-sectional area, the kinetic energy generated in passing through the restriction will be dissipated by the action of viscous forces and eddy formation. Hence at the downstream station 2 the velocity will be small and the difference between the kinetic energies of the fluid upstream and downstream of the restriction will be negligible. We thus see that the final specific enthalpy h_2 will be sensibly equal to the initial specific enthalpy h_1, but that the specific enthalpy will first fall and then rise again as the fluid flows from station 1 to station 2; consequently, the process must *not* be described as being *isenthalpic*. The overall process is highly irreversible; this aspect is discussed further in Chapter 9.

7.12 General-flow processes

In the present context, we shall use the term *general-flow process* simply to denote any flow process which does not fit into the two restricted categories of semi-flow and steady-flow processes with which we have just dealt.

As examples of general-flow processes, we might consider the following:

(a) Blowing up an expanding balloon. This would be a semi-flow type of process, but one in which we do not have a fixed control volume. Here we would have to take account of displacement work performed by the expanding control volume against the external pressure, and possibly also heat exchange between the control volume and its environment.

(b) Flow of feed water into a boiler drum while steam is drawn from it and, at the same time, the water level in the drum changes. Here the control volume would be fixed, but we would have to allow for the difference between the internal energy of the drum contents at the beginning and end of the period in question, and possibly also allow for heat transfer to the drum contents.

We do not need to go through either of these processes in any detail, for we can see by now that the Energy Conservation Equation for the control volume in question must satisfy the following general expression:

$$E_{c_i} + \text{Energy in} - \text{Energy out} = E_{c_f}. \tag{7.17}$$

It will then simply be a matter of taking proper account of, and correctly evaluating, all energy flows into and out of the control volume. We must therefore end on the warning note that the energy flow associated with mass flow can only be equated to the flow of enthalpy when the fluid is coming from or passing out to a region at uniform constant pressure; at the same time we must remember that it may also be necessary to take into account the kinetic energy of the fluid, and possibly also its potential energy.

7.13 Summary

In Chapter 5 we related energy change to adiabatic work, while in Chapter 6 we related the heat input to a system to the energy change of the system in a heat-only process. In the present chapter we first treated non-cyclic processes involving both work and heat interactions. In this way, we derived what is commonly called the *Energy Conservation Equation for a system*. Applying the result to a cyclic process, we then derived an expression which is frequently described as a statement of the First 'Law' of Thermodynamics, to which we gave the title *Cyclic Statement* of this 'Law' in order to distinguish it from the *Non-cyclic Statement* discussed in Chapter 4. We followed this with illustrations of the application of the Energy Conservation Equation to a number of simple non-flow processes.

Turning to flow processes, we used our *system analysis* of a non-flow process to derive an *Energy Conservation Equation for a control volume* and thereby showed how to apply *control-volume analysis* to a variety of flow processes. Here we encountered for the first time the energy-type property of a system which we call *enthalpy*. We paid particular attention to steady-flow processes and illustrated the

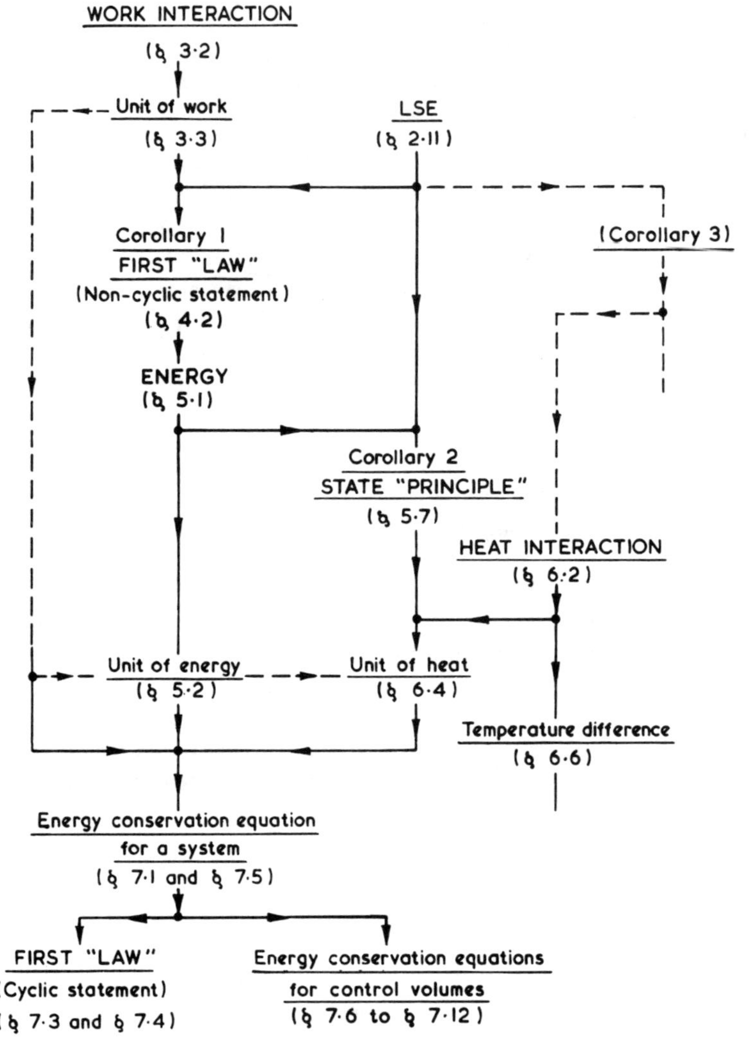

Fig. 7.6 Further development of the Thermodynamic Family Tree in Chapter 7

application of the *Steady-Flow Energy Equation* to a number of simple engineering processes. Our Thermodynamic Family Tree now appears as shown in Fig. 7.6.

The following Appendix A on the thermodynamic properties of pure substances is inserted here in order to provide material for the performance of illustrative calculations relating to the concepts so far presented.

APPENDIX A

Some Data on Pure Substances and Perfect Gases

In order to facilitate the calculation of problems relating to processes involving pure substances, this Appendix presents some useful data and diagrams relating to their thermodynamic properties. Further data and diagrams involving entropy are presented in Appendix E of Chapter 12. In Chapter 18 there is a more comprehensive treatment of the thermodynamic properties of Simple Systems, of which pure substances represent a particular case; indeed, a pure substance as we treat it here is a somewhat idealized concept, since substances tend to break down into mixtures of simpler chemical species to an increasing extent as their temperature rises. This is the phenomenon of *dissociation*, but for present purposes we may ignore this aspect of the matter.

In this preliminary study of some of the thermodynamic properties of pure substances, we shall be treating unit mass of the substance when it is at rest under the conditions specified in the definition of a Simple System given in Section 5.3.

A.1 Identifying the Stable State of a pure substance

In order to identify the Stable State of a pure substance, we need to specify the values of *only two* independent thermodynamic properties. In principle, any two will serve this purpose but some pairs are more convenient than others, as we shall discover in Chapter 18. We shall there also explain why it is sufficient to specify the values of only two.

That it is sufficient to specify only two implies that we may plot property diagrams for pure substances in x, y, and z cartesian coordinates, since we shall have $z = z(x,y)$, where x and y are the two chosen independent variables. These diagrams will thus be representations of *three-dimensional thermodynamic surfaces*.

A.2 $p-v-T$ diagram for a pure substance

Figure A.1 shows such a three-dimensional surface, on which are clearly marked those parts of the surface in which the substance exists in the solid, liquid, and gaseous phases respectively. Also marked are those parts of the surface respectively representing two-phase mixtures of either (a) solid and liquid, (b) solid and gas, or (c) liquid and gas. These areas are bounded by dotted lines which separate them

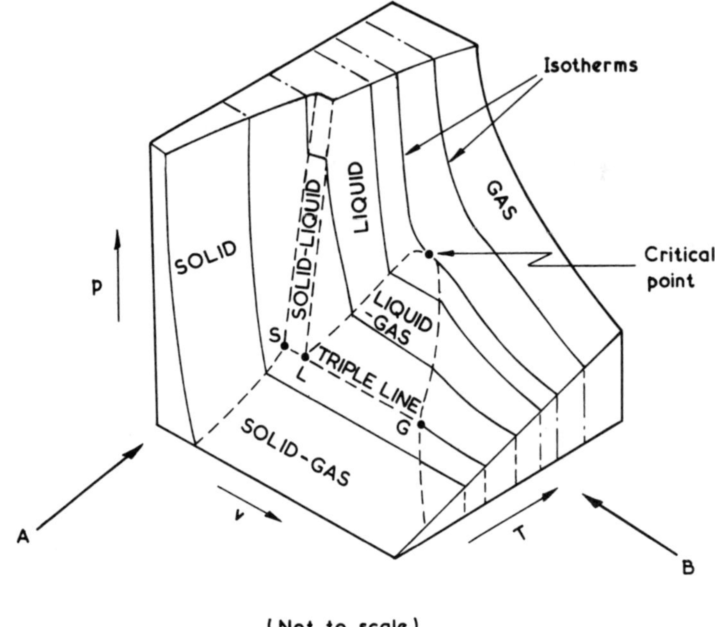

Fig. A.1 Three-dimensional $p-v-T$ surface for a pure substance that contracts on freezing

from the single-phase regions. The diagram is drawn for a substance which contracts on freezing, whereas water expands.

At the pressure and temperature represented by the *triple line* (SLG), the solid, liquid, and gas can coexist in equilibrium with each other in any proportions in a mixture of the three phases.

At temperatures greater than that at the *critical point*, there is no change from the gaseous phase to the liquid phase as the gas is compressed from a low pressure to a high pressure. It is clear from the diagram that no sharp dividing line can be drawn to separate that part of the three-dimensional $p-v-T$ surface which is labelled 'liquid' from that which is labelled 'gas'; the 'liquid' and 'gas' phases merge imperceptibly into each other as one traces out a path which lies wholly in the single-phase region.

The symbol T denotes *thermodynamic temperature*; we shall define this formally in Chapter 11.

A.3 $p-v$ diagram for a pure substance

If we view from direction A the three-dimensional $p-v-T$ surface depicted in Fig. A.1 and project this view on to a back plane, we obtain the $p-v$ diagram of Fig. A.2. The isotherms are here analogous to the contours on a geographical map. We may note the dotted lines representing the loci of state points for *saturated*

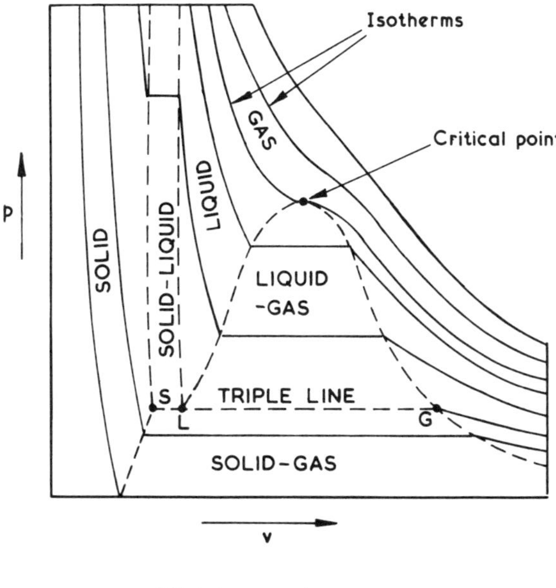

(Not to scale)

Fig. A.2 $p–v$ diagram for a pure substance that contracts on freezing, showing isotherms

solid, saturated liquid, and *saturated gas (vapour)* respectively; the term *saturated* here denotes the condition of the single pure phase (solid, liquid, or gas respectively) when it is in equilibrium with a mixture of either or both of the other two phases, the latter condition occurring only at the triple line.

A.4 $p–T$ diagram for a pure substance

If we view from direction B the three-dimensional $p–v–T$ surface depicted in Fig. A.1 and project this view on to a back plane, we obtain the $p–T$ diagram of Fig. A.3. We have here omitted the isotherms drawn on the two previous diagrams, so that all we see are the three *saturation loci,* which represent edge-on views of the three surfaces comprising the three two-phase regions, since all isotherms on these surfaces are horizontal in Fig. A.1. These three saturation loci intersect at the *triple point*, which is an end-on view of the triple line of Figs. A.1 and A.2.

A.5 Dryness fraction of a liquid–gas (vapour) mixture

In engineering, we encounter the liquid–gas two-phase region far more frequently than the other two-phase regions. We shall therefore study this region in more detail, while noting that similar considerations may be applied to the other two-phase regions.

In a two-phase mixture of saturated liquid and saturated gas (vapour), the

74

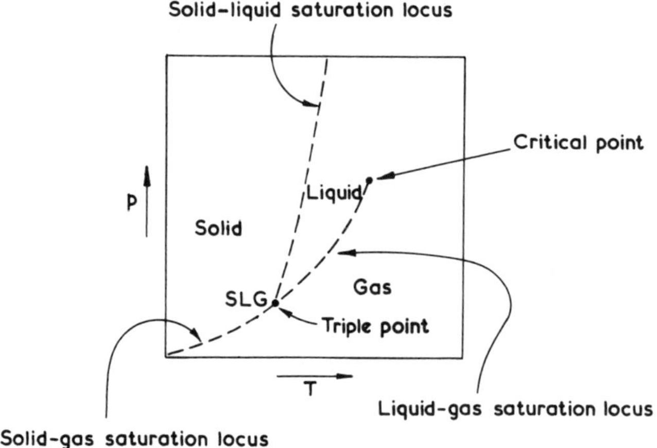

(Not to scale)

Fig. A.3 $p-T$ diagram for a pure substance

dryness fraction x of the mixture is defined as the mass fraction of gas present in the mixture. Thus, if a given quantity of the mixture contains a mass m_f of saturated liquid and a mass m_g of saturated gas, the dryness fraction of the mixture is defined as follows:

Definition

$$x \equiv \frac{m_g}{m_f + m_g}.$$
(A.1)

A line of constant dryness fraction x is shown in the two-phase region of Fig. A.4. In order to plot this line, we note that, if we have unit mass of mixture, then the mean specific volume of the mixture will be given by

$$v_C = x v_g + (1 - x) v_f,$$
(A.2)

where v_f and v_g are respectively the specific volume of the saturated liquid at A and the specific volume of the saturated gas at B. Thus, we shall have

$$x = \frac{v_C - v_f}{v_g - v_f} = \frac{AC}{AB}.$$
(A.3)

Hence, in order to plot a line of constant dryness fraction, we have only to set off this fraction on a series of isotherms (which are also isobars in the two-phase region) and join up the resulting points with the chain-dotted line shown in Fig. A.4. This line clearly passes through the critical point.

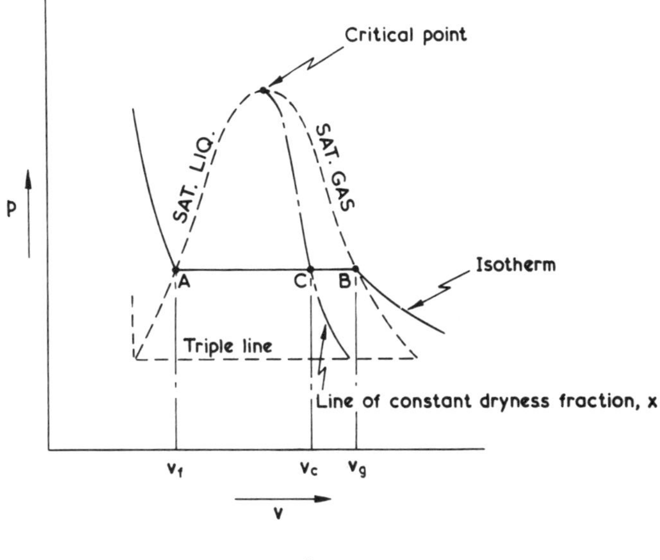

Fig. A.4 Liquid–gas, two-phase region on the $p-v$ diagram

A.6 $u-v$ diagram for a pure substance

In this chapter we have been concerned with calculations involving the internal energy and the enthalpy of a system, and we are now in a position to illustrate on separate diagrams the variations in the values of these properties for a pure substance.

We have already noted in Section A.1 that we may represent the relation between any three thermodynamic properties of a pure substance by a three-dimensional thermodynamic surface. We may then draw two-dimensional diagrams by projecting features of this surface on to a chosen plane and use the projections of contour lines to show how the third property varies with variations in the other two.

Without drawing it in full, we will now consider the $u-v-T$ three-dimensional surface for a pure substance by drawing a $u-v$ diagram on which we shall depict a few isotherms, as shown in Fig. A.5. We could, of course, have considered alternatively the $u-v-p$ surface.

The chief point of academic interest in the $u-v$ diagram of Fig. A.5 lies in the fact that, instead of a triple *point* as in Fig. A.3, and a triple *line* as in Figs. A.1, A.2, and A.4, we now have the triple *area* (SLGS) depicting the conditions under which the saturated solid, saturated liquid, and saturated vapour can coexist in equilibrium together in varying proportions. However, we shall rarely encounter these conditions in engineering applications.

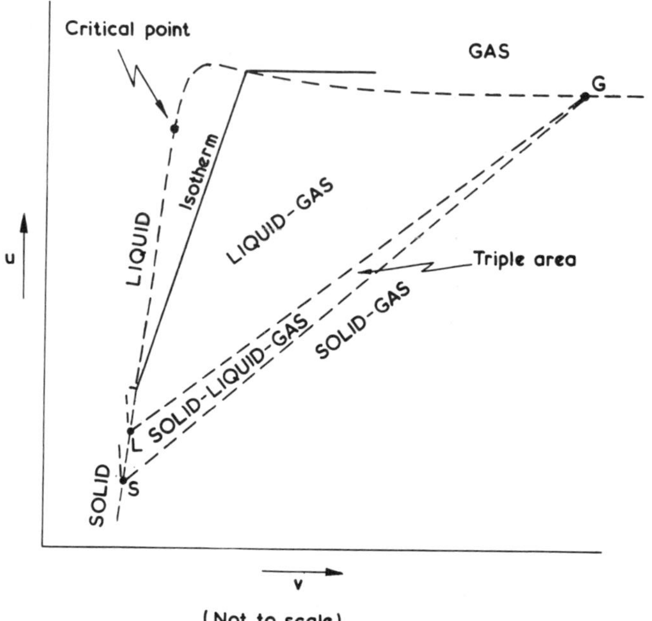

(Not to scale)

Fig. A.5 Internal energy–specific volume diagram for a pure substance which contracts on freezing

A.7 p–h diagram for a pure substance

A diagram which is particularly useful in refrigeration engineering is the pressure–enthalpy diagram for a pure substance. Fig. A.6 presents such a diagram, on which are shown a few isothermal contours, so that it represents the projection on the p–h plane of the p–h–T three-dimensional thermodynamic surface. A practical example of such a diagram[8] also depicts contours of constant specific volume v and constant specific entropy s (we shall not encounter this property until we reach Chapter 12); in this case, such a diagram represents the superposition, in one diagram, of individual projections on the p–h plane of the three separate thermodynamic surfaces p–h–T, p–h–v, and p–h–s.

In Fig. A.6, only the liquid and gaseous phases are shown on the diagram, since there is little engineering interest in the solid phase. It is usual to plot p on a logarithmic scale, since this proves to be more convenient. A line of constant dryness fraction x is shown on the diagram; as in the case of Fig. A.4, this line is constructed by noting that, for a typical isotherm,

$$x = \frac{AC}{AB},$$

since

$$x = \frac{h_C - h_f}{h_g - h_f}, \tag{A.4}$$

where h denotes specific enthalpy, namely the enthalpy per unit mass of substance.

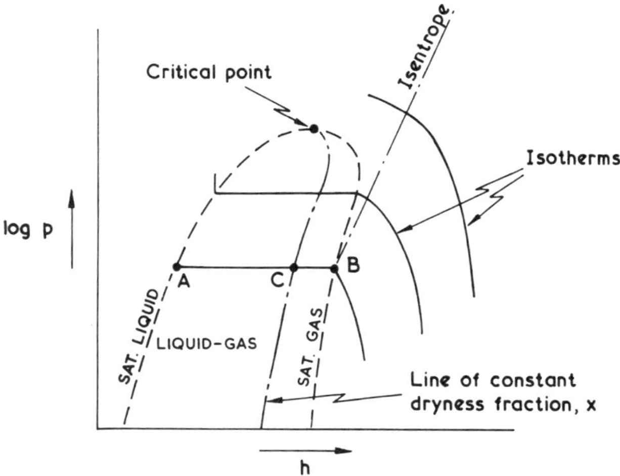

Fig. A.6 Pressure—enthalpy diagram for a pure substance

A.8 Specific heat capacities

We have two specific heat capacities of a pure substance: one at constant volume, c_v, and the other at constant pressure, c_p. These are defined by the following expressions:

Definitions

$$c_v \equiv \left(\frac{\partial u}{\partial T}\right)_v \tag{A.5}$$

$$c_p \equiv \left(\frac{\partial h}{\partial T}\right)_p \tag{A.6}$$

A.9 Ideal, perfect, and semi-perfect gases

We shall use the term *ideal* gas to encompass both *perfect* and *semi-perfect* gases. They are hypothetical gaseous substances to which the behaviour of real gases more closely approximates as the pressure falls towards zero. We define them formally as follows:

Definitions

An *ideal gas* is one whose p–v–T equation of state is

$$pv = RT, \tag{A.7}$$

in which T denotes *thermodynamic temperature*, so that $T = (t + 273.15)$ kelvins, where t is the Celsius temperature (see Chapter 11).

A *perfect gas* is an ideal gas whose specific heat capacities are constant.

A *semi-perfect gas* is an ideal gas whose specific heat capacities are not constant but are functions only of temperature.

78

Thus the $p-v-T$ equation of state of both perfect and semi-perfect gases is $pv = RT$. It can be shown mathematically (see Problem A.5) that, for any pure substance *having this equation of state, u, h, c_v and c_p are all functions only of temperature,* so that the definition of a semi-perfect gas is in accord with this fact. A perfect gas is thus a special case of a semi-perfect gas for which, for example, $c_v = c_v T^0$, since $T^0 = 1$.

A gas may be treated as perfect over a limited temperature range when at low pressure, namely at a pressure well below the critical. Over wider temperature ranges, such as are encountered in combustion processes, account must be taken of the increase in the values of the specific heat capacities with rise in temperature, so that we must then treat the gas as semi-perfect. At high pressures, no gas may be treated as ideal. We treat perfect gases next, but leave the treatment of semi-perfect gases to Chapter 17.

It should be noted that the equation of state of a substance is applicable only to states of stable equilibrium, namely Stable States.

A.10 Perfect gases

We noted above that, for any substance whose $p-v-T$ equation of state is $pv = RT$, the specific internal energy u and specific enthalpy h are functions only of T.* From this, it follows that, for *any* process undergone *by a perfect gas* between two states, *the following expressions are valid whether or not there is constancy of v or p during the process:*

$$du = c_v \, dT \tag{A.8a}$$

or

$$u_2 - u_1 = c_v(T_2 - T_1), \tag{A.8b}$$

and

$$dh = c_p \, dT \tag{A.9a}$$

or

$$h_2 - h_1 = c_p(T_2 - T_1). \tag{A.9b}$$

We may use equations (A.8a) and (A.9a) to obtain a useful relation between c_v, c_p, and R for a perfect gas. Thus

$$(c_p - c_v)dT = dh - du. \tag{A.10}$$

From the definition of enthalpy (Section 7.8) we have

$$dh = du + d(pv),$$

---

* That this behaviour is closely approached by real gases at low pressures is evident from the fact that, as the pressure falls, the isotherms tend to become horizontal in Fig. A.2 and vertical in Fig. A.6.

which, from equation (A.7) and noting that R is constant, gives

$$dh - du = d(pv) = d(RT) = R\,dT. \tag{A.11}$$

Hence, from equations (A.10) and (A.11) we obtain the relation

$$\boxed{c_p - c_v = R} \tag{A.12}$$

which, from the method of its derivation, is seen to apply not only to perfect gases but to all ideal gases (i.e. gases which have the equation of state $pv = RT$). We shall find this relation useful when deriving one of the expressions for the change in entropy of a perfect gas in Appendix E of Chapter 12.

Since c_v and c_p are both constant for a perfect gas, so is their ratio, to which we give the symbol γ, so that

$$\gamma \equiv \frac{c_p}{c_v}. \tag{A.13}$$

Hence, from equation (A.12) we have

$$c_v = \frac{1}{\gamma - 1} R \tag{A.14}$$

and

$$c_p = \frac{\gamma}{\gamma - 1} R. \tag{A.15}$$

A.11 Further data on pure substances

Further data on the thermodynamic properties of pure substances, including diagrams relating to entropy, are given in Appendix E of Chapter 12, while a more comprehensive treatment appears in Chapter 18.

CHAPTER 8

The Second 'Law' of Thermodynamics (With Appendix B)

8.1 Introduction

In Sections 4.2 and 7.4, it was shown that propositions which have traditionally been elevated to the status of a Law under the title of the First Law of Thermodynamics can in fact be derived as corollaries of the Law of Stable Equilibrium (Section 2.11). For that reason, the word 'Law' was placed within inverted commas in the title of Chapter 4. For similar reasons this word has also been placed within inverted commas in the title of the present chapter.

While the Non-cyclic Statement of the First 'Law', namely Corollary 1 of the Law of Stable Equilibrium, led to the definition of the thermodynamic property E which we call *energy*, we shall find that the proposition which is traditionally associated with the title Second 'Law', and which in this chapter we shall prove as a further corollary of the LSE, will lead in Chapters 11 and 12 to definitions of two further thermodynamic properties, namely *thermodynamic temperature T* and *entropy S* respectively.

As we have seen, First 'Law' considerations are closely related to the notion of energy conservation and so provide us with means for devising energy accounting systems in the form of *energy balances* for our working devices. These simply allow us to indicate what happens to our energy inputs and outputs when we have built a device, but they tell us nothing about the measure of excellence of the device, as compared with hypothetical ideal devices. On the other hand, it is a very important feature of Second 'Law' considerations that they do lead to propositions which enable us to set up *performance criteria* for our engineering devices. These propositions must await development in a later chapter; in this chapter we commence our study of Second 'Law' considerations by proving two further corollaries of the Law of Stable Equilibrium. The second of these corollaries constitutes, in fact, one of the conventional statements of the Second 'Law'.

8.2 Corollary 3 of the LSE – The impossibility of constructing a Non-cyclic PMM 2 (Non-cyclic Statement of the Second 'Law')

In establishing the impossibility of constructing the hypothetical *non-cyclic* device which we here describe as a *Non-cyclic Perpetual Motion Machine of the Second Kind* (a *Non-cyclic PMM 2*, for short), we shall be meeting for the first time some ideas associated with the Second 'Law' of Thermodynamics. It is for this reason that we refer to the device as being of the 'Second Kind'. (Historically, it may be of interest to note that the name 'Perpetual Motion Machine of the First Kind' has been given to a hypothetical *cyclic* device associated with First 'Law' considerations; this is one in which a system produces positive net work output while undergoing a completed, adiabatic cyclic process. The reader should have little difficulty in establishing the impossibility of constructing such a device.)

Before proceeding to a formal definition of a Non-cyclic PMM 2, it is well to recall that, in Section 2.5, we gave the name *Constrained System* to one which is subject to specified Constraints, including confinement within a fixed bounding surface, and for which none of the Constraints to which it is subject are altered during the process under consideration; in the same Section, we also gave the name *Allowed States* to all those states which, through interaction with the environment or otherwise, are potentially accessible to the system in these circumstances. Bearing this in mind, we now define a Non-cyclic PMM 2 in the following terms:

Definition

Any hypothetical device by means of which a Constrained System could be taken from *an initial Stable State* to another Allowed State in a non-cyclic process, on completion of which the sole end effect external to the system could have been the change in level of a weight to a *higher** position, is called a *Non-cyclic Perpetual Motion Machine of the Second Kind* (a *non-cyclic PMM 2*, for short).

Diagramatically, we may thus represent this hypothetical device in the manner depicted in Fig. 8.1. The reason for calling it a Perpetual Motion Machine will be discussed in Section 8.4.

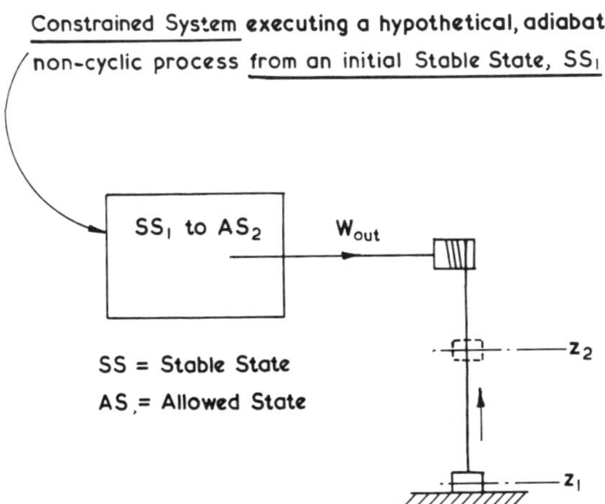

Fig. 8.1 Diagrammatic representation of a Non-cyclic PMM 2. [*Note:* The term Constrained System (Section 2.5) denotes that the Constraints to which the system is subject *remain unaltered.*]

* The presence of a gravitational field is presupposed.

82

A Non-cyclic PMM 2 is defined in terms of a Constrained System because, as we recall from Section 2.11, the Law of Stable Equilibrium relates to Constrained Systems and it is this Law that postulates the possibility of existence of Stable States. We shall use the definition of a Stable State given in Section 2.9 to prove the following theorem, which we may describe as a corollary of the Law of Stable Equilibrium since it is the LSE which tells us that, while remaining isolated from its environment, a Constrained System will always ultimately take up a Stable State:

Corollary 3 of the LSE

It is impossible to construct a Non-cyclic PMM 2.

To prove this corollary, we first recall from the discussion in Section 2.9 that, when the Constrained System is in its initial Stable State, the individual macroscopic parts into which the system may conceptually be divided are in mutual equilibrium with each other. Now let us suppose that, contrary to Corollary 3 and as indicated in Fig. 8.1, a Constrained System executes a hypothetical, non-cyclic process from an initial Stable State SS_1 to an Allowed State AS_2, on conclusion of which the sole end effect external to the system has been the raising of a weight (i.e. the production of positive net work output). We could then allow the weight to descend to its original level while feeding work locally into *part* of the system (e.g. by means of a thermally non-conducting paddle or friction wheel), thereby resulting temporarily in a non-equilibrium state of the system at the instant when the weight was restored to its original level. Alternatively, the restoration of the weight to its initial position could have been used to give the system a finite velocity. Consequently, the overall process would have been one in which the system had been taken from an initial Stable State to some other Allowed State AS_3, yet in which there had been no net effect on the environment. The possibility of such a course of events is ruled out by the very definition of a Stable State given in Section 2.9. Corollary 3 is thus proved.

It is evident that a device which brought about the inverse paddle-wheel or stirring process would constitute a Non-cyclic PMM 2, so that in proving Corollary 3 we have established, as was forecast in Section 4.2, that it is impossible to execute such a process.

8.3 Examples of possible and impossible processes
under the terms of Corollary 3

It is important to have a clear understanding of the implications of Corollary 3. Moreover, it is just as necessary to appreciate which types of process are not prohibited by Corollary 3 as it is to recognize those that are. To this end, the reader is invited to consider whether the following processes would contravene Corollary 3 and to set down careful explanations in support. Answers are given in Appendix B at the end of this chapter, but these should only be consulted when the exercise has been completed.

(a) A piston is free to move within a closed cylinder whose walls are rigid and non-conducting. At a certain instant, the steady pressures on the two sides of the piston are p_0 and p_1, where $p_1 > p_0$. As the piston moves it is caused to raise a weight external to the cylinder.

(b) Through the action of a restraining pin, a piston is held in a certain fixed position within a closed cylinder whose walls are rigid and non-conducting, the steady pressures on the two sides of the piston being p_0 and p_1, where $p_1 > p_0$. The pin is withdrawn and as the piston moves it is caused to raise a weight external to the cylinder.

(c) Through the action of a restraining pin, a piston is held in a certain position within a cylinder whose walls are rigid and non-conducting, the back of the piston being open to the atmosphere at pressure p_0 and the pressure within the cylinder being steady at p_1, where $p_1 > p_0$. The piston is caused to move inwards through the application to it of an externally applied force.

(d) A vessel with rigid, non-conducting walls is in the form of a cylinder with its axis horizontal. It contains a fluid in which there is initially a high degree of swirl. Through the action of a paddle wheel placed in the fluid an external weight is caused to rise as the degree of swirl is reduced.

(e) A vessel with rigid, non-conducting walls is divided by a rigid, non-conducting partition pierced by a central circular passage. Within the passage there is mounted a small turbo-rotor mounted on a non-conducting shaft which passes through a leak-proof gland in one wall of the vessel. The passage is sealed by a thin, non-conducting diaphragm ahead of the rotor. The two halves of the vessel contain gas, the pressures ahead of and behind the diaphragm being respectively p_1 and p_0, where $p_1 > p_0$. The complete system is initially in a Stable State. The diaphragm bursts and the resulting flow of gas causes the rotor to turn and to raise a weight external to the vessel.

(f) A fluid is contained within a rigid, non-conducting vessel and is initially in a Stable State. The temperature and pressure of the fluid change as the result of (1) the raising and (2) the lowering of a weight attached by a cord to a paddle wheel placed within the fluid.

(g) A non-conducting piston is free to move within a closed cylinder whose walls are rigid. With the exception of one end wall, the walls are also non-conducting. The system comprising the piston and the two fluids is initially in a Stable State. Heat is then supplied to the uninsulated end of the cylinder from a Constrained System in an initial Stable State, causing the piston to move and to raise a weight external to the cylinder.

(h) A fluid is contained within a rigid vessel and is initially in a Stable State. Heat is transferred from the vessel to a cyclic heat power plant (as defined in Section 8.5 hereafter), as the result of which a weight is raised, with no other effect external to the complete system.

It should be noted that a non-conducting wall is *not* a Constraint in the sense in which we have defined that term in Section 2.5. It is introduced in some of the above examples simply in order to provide the possibility only of a work interaction through it.

8.4 The significance of a Non-cyclic PMM 2

We may note from Examples (b), (e), and (f) of Section 8.3 that, if a system is in a Stable State, then, in an *adiabatic* process starting from that Stable State, positive work output can only be obtained if a Constraint, either external or internal, is relaxed; the system is not then a Constrained System in the sense defined in Section 8.2. However, we also see from Example (g) that, if the process is not adiabatic, then it is possible to obtain positive work output from an appropriate Constrained System starting from an initial Stable State.

From Example (f) we reach the important conclusion (forecast in Section 4.2) that, whereas the adiabatic stirring or *paddle-wheel process* of Fig. 8.2(a) is possible (as we already knew from practical experience), our proof of Corollary 3 has demonstrated the impossibility of bringing about the inverse paddle-wheel process of Fig. 8.2(b). This conclusion is of particular interest to us because the paddle-wheel process of Fig. 8.2(a) is an important example of an *irreversible* process, a term which bears a special connotation in thermodynamics. The source of the irreversibility lies not in the initial input of work from the paddle to the fluid but,

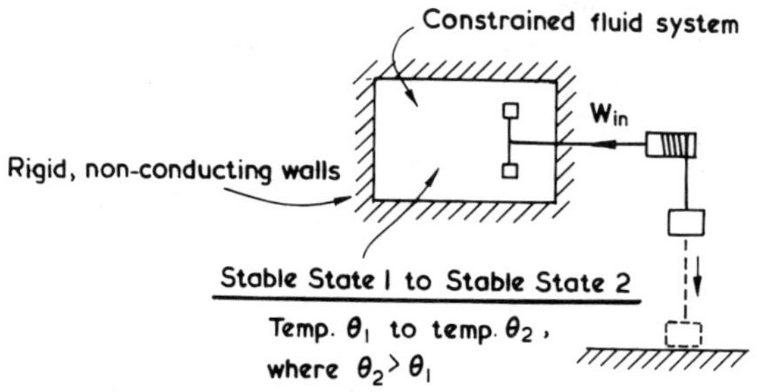

(a) Paddle-wheel process (possible)

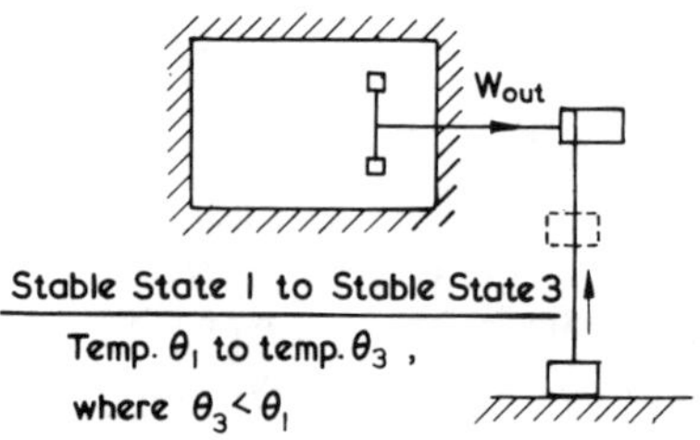

(b) Inverse paddle-wheel process (impossible - a Non-cyclic PMM 2)

Fig. 8.2 The paddle-wheel and inverse paddle-wheel processes

as was noted in Section 2.14, in the subsequent settling down of the fluid from a non-equilibrium state to a Stable State as the eddies so created finally subside, that settling down being a one-way process. We shall undertake a more detailed study of irreversible processes and their effects in the next and subsequent chapters.

The name *Non-cyclic PMM 2* has not hitherto been given to the hypothetical device here given that name, but the proof of the impossibility of constructing such a device was first given, in effect, by Hatsopoulos and Keenan.* We call it a 'Perpetual Motion Machine' because, if it were possible to create one, we could obtain directly from the oceans, for example, an almost unlimited amount of work: we could, in conception, isolate them and then, from their initial Stable State, cause work to be produced in an unaided manner so long as this hypothetical process could proceed. That would be a very long time indeed. A further reason for giving this hypothetical device the title of a PMM 2 arises from the fact that it would permit of the construction of a cyclic device which has long been known simply as a PMM 2 but which, to distinguish it from the device of Corollary 3, we shall call a *Cyclic PMM 2*. It is again a hypothetical device which it is impossible to construct.

Before demonstrating the impossibility of constructing a Cyclic PMM 2, we introduce a term which we shall frequently encounter in our later studies of thermodynamics.

8.5 Cyclic Heat Power Plant (CHPP)

We shall describe as a *Cyclic Heat Power Plant* (*CHPP*, for short) any device in which, in either a non-flow or flow process, a system or fluid is taken from an initial Stable State round a completed cyclic process in the course of which it exchanges heat with other systems, while the device produces a positive quantity of net work output as the result of delivering work at some points in the cyclic process and absorbing work at others. (We may recall from Section 7.2 that the need to specify an initial Stable State arises simply from the fact that classical thermodynamics treats matter as a continuum, so that it is not meaningful to talk about a cyclic process starting from a state which is not a Stable State.) An example of such a device is provided by a simple cyclic steam power plant, with heat absorption and rejection respectively in the boiler and condenser, work production in the turbine, and work absorption in the feed pump. Unfortunately, such devices are frequently called 'heat engines'. We shall avoid the use of this term since, in the example just cited, the 'heat engine' comprises all four components, namely boiler, turbine, condenser, and feed pump.

In general, when a system is made to undergo such a cyclic process, it may, subject to one important proviso, exchange heat with any number of other systems,

---

* It was given in their proof, in reference 1, of the following proposition: *No possible variation from a normal stable state can have as sole external effect the rise of a weight*. In the proof, they used the raised weight to give the system a finite velocity as the weight was lowered, so leaving the system in a state different from the initial Stable State after the weight had been restored to its initial position. See also the Corollary on page 35 of reference 1 and Corollary C on page 376.

the proviso being that it is not possible to construct a CHPP in which, during a completed cycle, the net work output from it is equal to the heat supplied to it during the cycle, with therefore no heat rejected by it. Such a CHPP would, in effect, exchange heat with only a single system and would constitute a Cyclic PMM 2, as defined in the next Section. The proposition that it is impossible to construct such a device is presented in many texts as an unproven statement to which is given the title of the Second Law of Thermodynamics. We shall demonstrate, however, that the proposition follows directly from Corollary 3 of the Law of Stable Equilibrium established in Section 8.2; we shall call this proposition the *Cyclic Statement of the Second 'Law'*.

8.6 The impossibility of constructing a Cyclic PMM 2
(Cyclic Statement of the Second 'Law')

We define a Cyclic PMM 2 formally as follows:

Definition

Any hypothetical device in which, during a completed cycle, a Cyclic Heat Power Plant (CHPP) delivers a net work output equal to the heat supplied to it during the cycle, with therefore no heat rejected by it, is called a *Cyclic Perpetual Motion Machine of the Second Kind* (a *Cyclic PMM 2*, for short).

The hypothetical CHPP shown within Fig. 8.3 represents such a Cyclic PMM 2. We shall now make use of Corollary 3 of the Law of Stable Equilibrium to prove the following proposition:

Proposition

The *Second 'Law' (Cyclic Statement)* states that *it is impossible to construct a Cyclic PMM 2*.

We may demonstrate the truth of this Proposition by using as our source of heat supply to a hypothetical Cyclic PMM 2 a Constrained System which is initially in a Stable State, in the manner indicated in Fig. 8.3.

According to Corollary 3 of Section 8.2, it is not possible to construct a device (a Non-cyclic PMM 2) in which a Constrained System is taken from an initial Stable State to some other Allowed State with the sole net end effect external to the system being the raising of a weight. Now the hypothetical CHPP within Fig. 8.3 causes the Constrained System within control volume X to be taken through just such a process, for the net work output produced by the hypothetical CHPP in a completed cycle could have been used to raise a weight; moreover, this would have been the sole net end effect external to the Constrained System, since the CHPP constitutes a cyclically operated system whose state remains unchanged at the end of a completed cycle. It is therefore not possible to construct a device such as that represented within control volume Y of Fig. 8.3, for this would constitute a Non-cyclic PMM 2 (albeit containing a cyclic device within itself). However, we know that it is possible for a Constrained System to be taken from an initial Stable State to some other Allowed State by heat transfer to or from it; hence this part of

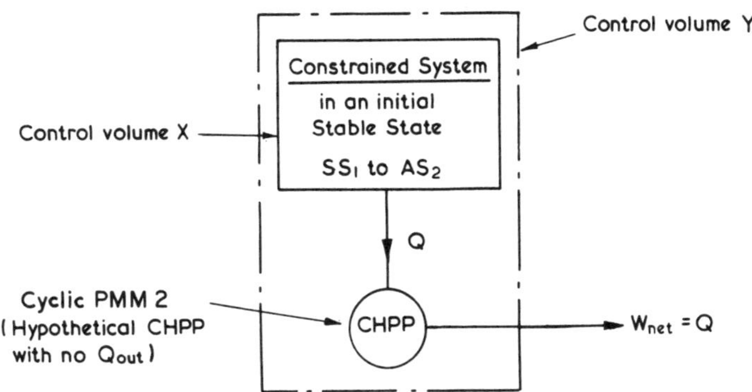

Fig. 8.3 The impossibility of realisation of a hypothetical Cyclic PMM 2

the complete process represented in Fig. 8.3 is not impossible of realisation. It must therefore be the hypothetical Cyclic PMM 2 that is impossible of realisation. Our Proposition is thus proved. The term PMM 2 thus covers both non-cyclic and cyclic Perpetual Motion Machines of the Second Kind.

We may finally point out that a real CHPP, in which there is heat exchange with both a source (System 1) and a sink (System 2) as represented diagrammatically in Fig. 8.4, does not enable Corollary 3 to be contravened, since the effect external to System 1 is then not solely the production of positive net work output but also a change in the state of System 2.

Corollary 3 would also not be contravened by a work-*absorbing* cyclic device [such as that depicted in Fig. 10.6(a)] in which the net work *input* to the device during a completed cycle was equal to the heat *rejected* by it during the cycle. Work supply to a system conceptually involves a *falling* weight, not the raising of a weight. Such a cyclic process could, in fact, be executed by causing the state of a fluid within a rigid vessel to be changed by work input from a paddle wheel rotated within the fluid, and then restoring the system to its original state by transferring from it an amount of heat equal to the work put into the fluid by the paddle wheel.

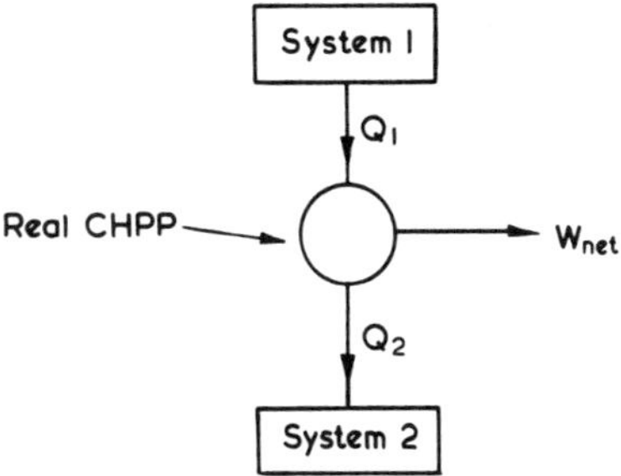

Fig. 8.4 Diagrammatic representation of a real CHPP, such as a simple cyclic steam power plant

8.7 Origin of the name 'Cyclic PMM 2'

The reason for calling a Cyclic PMM 2 a 'Perpetual Motion Machine' is essentially the same as that given in Section 8.4 for a Non-cyclic PMM 2, but in relation to indirect, instead of direct, production of work. It arises from the fact that, if a Cyclic PMM 2 were to exist, we could evidently obtain an almost unlimited amount of work by utilizing the oceans, for example, not as a direct source of work, as in a Non-cyclic PMM 2, but as a source of heat supply to a cyclic heat power plant which produced an amount of work equal to the quantity of heat supplied to it.

8.8 Alternative statements of the Second 'Law'

The statement that it is impossible to construct a Cyclic PMM 2 has for a long time been one of the forms in which the Second 'Law' of Thermodynamics has been quoted. Following this convention, we have therefore described the following two propositions as alternative statements of this 'Law'.

Propositions

The *Non-cyclic Statement of the Second 'Law'* states that *it is impossible to construct a Non-cyclic PMM 2.* (Corollary 3 of the LSE.)
The *Cyclic Statement of the Second 'Law'* states that *it is impossible to construct a Cyclic PMM 2.*

We showed that the truth of the first of these propositions followed as a corollary of the Law of Stable Equilibrium. We then used the first to establish the truth of the second. The use of the term 'Law' in relation to these two statements consequently no longer seems appropriate, for both propositions follow logically from the more fundamental statement known as the Law of Stable Equilibrium.

8.9 The thermal efficiency of a Cyclic Heat Power Plant (CHPP)

The impossibility of constructing a Cyclic PMM 2 tells us something of great importance to the engineer with respect to what has come conventionally to be described as the *thermal* or *cycle efficiency*, η_{CY}, of a Cyclic Heat Power Plant. This is defined formally by the expression

$$\eta_{CY} \equiv \frac{W_{net}}{Q_{in}}, \tag{8.1}$$

where W_{net} is the positive net work output produced during a completed cycle and Q_{in} is the heat supplied to the CHPP during the completed cycle. With respect to the CHPP depicted in Fig. 8.4, since $W_{net} = Q_1 - Q_2$, the cycle efficiency will be given by

$$\eta_{CY} = \left(1 - \frac{Q_2}{Q_1}\right). \tag{8.2}$$

It is evident from inspection of Figs. 8.3 and 8.4 that η_{CY} can never attain a value of 100 per cent., since Q_2 in Fig. 8.4 can never be zero. Before we can reach any conclusion about the factors that determine the actual value of η_{CY} in a given situation we shall have to continue our abstract thermodynamic studies and, in particular, await the definition in Chapter 11 of the property of a system which we call *thermodynamic temperature*. We shall then find that the magnitude of Q_2 in Fig. 8.4, relative to that of Q_1, is dependent on the thermodynamic temperatures of systems 1 and 2.

8.10 Summary

In this chapter, in the form of Corollary 3, we have established a third branch stemming from the head of the Thermodynamic Family Tree (namely from the Law of Stable Equilibrium). The first branch (Corollary 1) gave us in Chapter 4 the conventional non-cyclic statement of the First 'Law'. The second branch (Corollary 2) established the State 'Principle' in Chapter 5. These two branches combined together in Chapter 7 to yield the conventional 'Energy Conservation Equation' for a system, which enabled us then to deduce the conventional cyclic statement of the First 'Law' (which, it is of interest to note, many other texts take as their starting point in their presentation of classical thermodynamics).

Corollary 3 demonstrated the impossibility of constructing a *Non-cyclic Perpetual Motion Machine of the Second Kind* (a *Non-cyclic PMM 2*) and so the impossibility (foreseen in Section 6.2) of achieving the inverse paddle-wheel (stirring) process starting from a Stable State. The latter conclusion recalled to us the concept of thermodynamic irreversibility, which will be developed more fully in the next chapter. It was also pointed out that this hypothetical device was given this name because, if it were to exist, we could obtain directly from the oceans, for example, an almost unlimited amount of work.

Having defined what we mean by a *Cyclic Heat Power Plant* (commonly described misleadingly as a 'Heat Engine'), we then used Corollary 3 to demonstrate the impossibility of constructing a *Cyclic PMM 2*. This was defined as a hypothetical device in which, during a completed cycle, a Cyclic Heat Power Plant (CHPP) delivers a net work output equal to the heat supplied to it during the cycle, with therefore no heat rejected by it. We saw that this hypothetical device is described as a 'Perpetual Motion Machine' because, if it were to exist, we could obtain an almost unlimited amount of work by utilizing the oceans, for example, not as a direct source of work, as in a Non-cyclic PMM 2, but as a source of heat supply to a Cyclic PMM 2.

The proposition that it is impossible to construct a Cyclic PMM 2 has long been presented in many texts, without proof, as the *Second Law of Thermodynamics*. This led us to describe the proposition as the *Cyclic Statement* of this 'Law', while describing Corollary 3 as the *Non-cyclic Statement*. At the same time, we recognized the inappropriateness of using the term 'Law' in relation to either, since both have been seen to follow logically as corollaries of the more fundamental Law of Stable Equilibrium.

We ended the chapter by noting that, in consequence of the impossibility of constructing a Cyclic PMM 2, the *thermal efficiency* of a *Cyclic* Heat Power Plant (CHPP) can never attain a value of 100 per cent. The factors that determine the

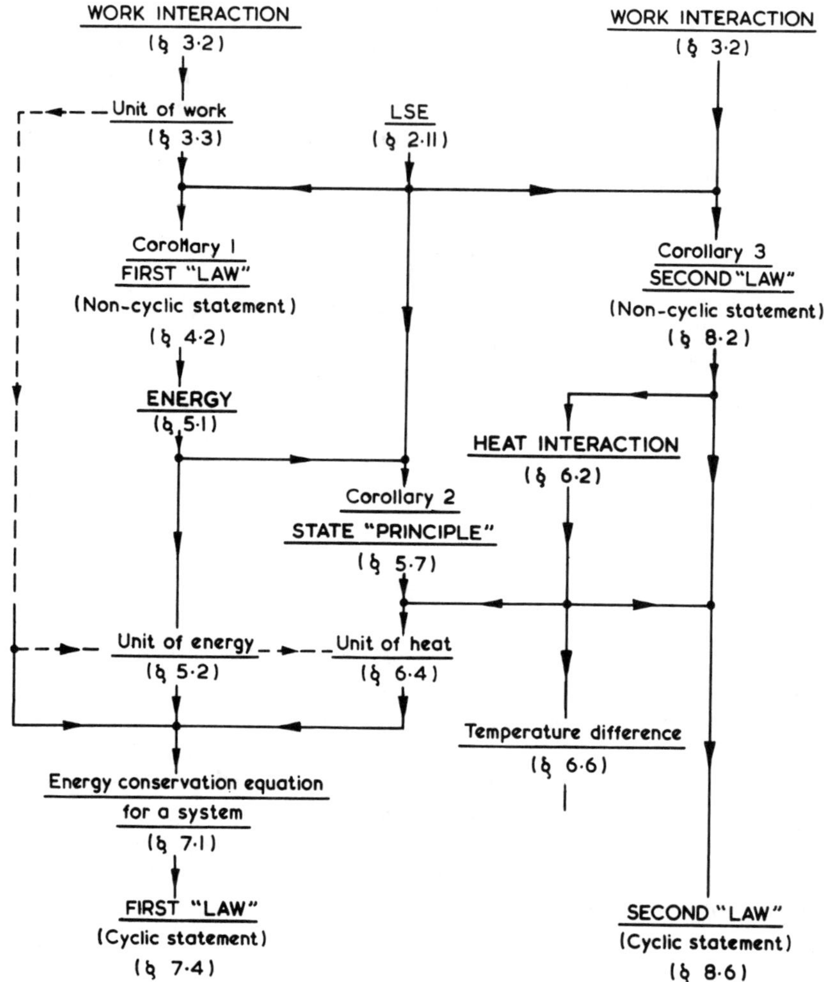

Fig. 8.5 Further development of the Thermodynamic Family Tree in Chapter 8

actual value of the thermal efficiency in a given situation will be studied in later chapters.

Our Thermodynamic Family Tree now appears as in Fig. 8.5, which is of interest in showing that, as a logical development from the Law of Stable Equilibrium, we have now arrived at the conventional cyclic statements of the First and Second 'Laws'. Our further extension of the Thermodynamic Family Tree will depict the application of these propositions to the development of the concepts of two important properties of a system: the *thermodynamic temperature* and the *entropy*.

APPENDIX B

Answers to the Examples of Section 8.3

'Yes' and 'No' mean respectively that the process *does* and *does not* contravene Corollary 3.

(a) *No.* Taking *the two fluids and piston together* as our system, although we have a Constrained System passing from one state to another Allowed State, the initial state was not a Stable State.

(b) *No. Taking the two fluids and piston together* as our system, we may look at the situation in either one of two different ways, namely:

Either

If we take State 1 as the state of the system before the pin was withdrawn, then although this was a Stable State our system was not a Constrained System (one for which there is no alteration of the Constraints during the process), since State 2 was reached only after relaxation of a Constraint (through withdrawal of the pin).

Or

If we take State 1 as the state of the system immediately after withdrawal of the pin, then although we have a Constrained System throughout the subsequent process, State 1 was not a Stable State (even though the properties of the system would have the same values *immediately* following the withdrawal of the pin as they did before; that situation would not last for any finite time).

(c) *No.* Taking *the fluid within the cylinder* as our system, this was not confined within a fixed bounding surface and so was not a Constrained System in the sense defined in Section 8.2. Moreover, work was put *into* the system, so that the end effect external to the system could have included the change in level of a weight to a *lower* position, not to a higher position.

(d) *No.* Taking *the fluid* as our system, although we have a Constrained System going from an initial state to another Allowed State, and the sole external effect was the raising of a weight, the initial state of the swirling fluid was not a Stable State (in time, the swirling would have died down of its own accord through the action of viscous forces in the fluid).

(e) *No.* Taking *the complete contents of the vessel* as our system, although this was initially in a Stable State, the bursting of the diaphragm constituted relaxation of an internal Constraint, so that Corollary 3 is not applicable, since this only applies to a system which remains subject to fixed Constraints.

92

(f) (1) *Yes.* Taking *the fluid* as our system,
we have a Constrained System being taken from an initial Stable State to an Allowed State, with the sole end effect external to the system being the *raising* of a weight. We thus have a Non-cyclic PMM 2.

(2) *No.* Taking *the fluid* as our system,
the sole end effect external to the system was the *lowering* of a weight; this is not forbidden by Corollary 3.

(g) *No.* Taking *the two fluids and piston together* as our system, this is a Constrained System which was in an initial Stable State, but the raising of a weight was not the sole end effect; heat was supplied to the system, which would have left a change in the state of the environment.

(h) *Yes.* (However, the answer is not directly obvious but depends on the knowledge, yet to be gained in Section 8.6, that a cyclic heat power plant, as there defined, must also reject some heat in addition to producing some work. In Section 8.6 we in fact use Corollary 3 to *prove* that this must be the case; i.e. we prove that it is not possible to construct what is there described as a *Cyclic* PMM 2. Note that both (g) and (h) involve heat exchange with only a single Constrained System in an initial Stable State, but that, whereas the work-producing device in (g) is non-cyclic, that in (h) is cyclic.)

CHAPTER 9

Irreversibility and Reversibility

9.1 Introduction

The concepts of thermodynamic irreversibility and reversibility have already received preliminary mention in the following contexts: (a) in Section 2.14 in the context of the Law of Stable Equilibrium, when we noted the impossibility of an isolated system departing from a Stable State once it had settled to that state; (b) in Section 6.8 in the context of the irreversibility of heat transfer between two systems of finitely different temperature; and (c) in Section 8.4 in the context of Corollary 3 of the Law of Stable Equilibrium and the consequent impossibility of bringing about the inverse paddle-wheel (stirring) process.

All these cases involve departures from equilibrium; this is the essential ingredient of all irreversible processes. Indeed, we can already say that if a process involves a finite departure from a state of stable equilibrium (i.e. from a Stable State), then irreversibility will be introduced into the process; furthermore, since in the real physical world all processes require to be carried out in a finite time, *all real-life processes are in some measure irreversible.* In this context, it is of interest to note that this italicized statement has been quoted[9] as yet another statement of the so-called Second 'Law'. We have already encountered it in Section 2.10 in our discussion of the internal condition of a system when it is in a Stable State, and from that discussion we can appreciate why the Law of Stable Equilibrium encompasses this alternative statement of the Second 'Law'.

Thus real processes only approach complete reversibility as a limiting ideal which exists only in that idyllic land of the thermodynamicist to which the author has elsewhere[10] given the name 'Thermotopia'; yet it is largely with the aid of calculations relating to hypothetical reversible processes that engineers and scientists are able to study and predict the actual behaviour of real systems and plants.

As was noted in Section 2.14, the great importance to the engineer of a study of the irreversibility of real physical processes lies in the fact that irreversibility always results in lost opportunities for producing work or in a greater work input than is ideally needed. The theoretical basis for this statement is established in the next Chapter.

9.2 Irreversible processes

In the three cases quoted above, the source of irreversibility has been seen to lie in departures from equilibrium. Further consideration of these cases reveals the

94

additional fact that, when a system undergoes an irreversible process, the combined effects of that process on the system and its environment can never be completely effaced. For example, in the first case (Section 2.14) of an isolated system settling to a Stable State from a non-equilibrium Allowed State, whereas it would theoretically be possible to bring the system back to its initial Allowed State, it would require the input of work from the environment to achieve this (e.g. if the system were a fluid, by the rotation of a paddle wheel within the fluid); the environment would then be left in a state different from its initial state, for it was unaffected when the system remained isolated during the original process. It will be a useful exercise for the reader to arrive at a similar conclusion in each of the other two cases (Sections 6.8 and 8.4). We may therefore usefully define an irreversible process as follows:

Definition

A process undergone by a system is *irreversible* if all the effects of the process on the system and its environment cannot be effaced.

It should be particularly noted that this definition does *not* refer to *reversal* of the process, although it is true that, when a process is irreversible in this thermodynamic sense, it is never possible to execute exact reversal of the process in every precise detail.

9.3 Test for irreversibility

We have noted in Section 9.1 that all real-life processes are in some measure irreversible and it might therefore be contended that it would be superfluous to devise a test for irreversibility. It is, however, useful to be able to categorize certain classes of process as being essentially irreversible and therefore to devise a test by means of which they can clearly be established as such. To this end, we use the above definition as our guide and the impossibility of constructing either a Non-cyclic or a Cyclic PMM 2 as our weapon; if we can show that the hypothetical effacing of the process under consideration would enable us to construct either of these hypothetical devices, then we know that the process itself cannot be effaced and is therefore irreversible. It is advisable to carry out this procedure methodically in the following series of steps:

(a) Sketch the process in its essential physical detail.
(b) Sketch the 'Effacer', namely the hypothetical process which would be needed to efface all the effects of the original process on both the system and its environment. Note particularly that this is *not* the reverse process (which, in fact, it is not possible to execute if the original process is irreversible), but the imaginary hypothetical process that would be needed to achieve complete effacement.

(c) Decide whether the Effacer itself constitutes either a Non-cyclic or Cyclic PMM 2. (If it does, the task is ended.)

(d) If the Effacer does not itself constitute a PMM 2, examine whether, with the aid of the Effacer and of real known physical processes, it is possible to construct either of these. If it is established that this is possible, then it cannot be possible for the Effacer to exist and the original process must therefore have been irreversible.

9.4 An example of testing for irreversibility

As an example of this procedure, we demonstrate that the steady-flow *adiabatic throttling* process, illustrated as process (c) in Fig. 7.5, is irreversible. We do this in the following three steps, shown in Fig. 9.1:

(a) This depicts an adiabatic throttling process.

(b) This depicts the Effacer.

(c) This demonstrates that, with the aid of the Effacer and other real known physical processes, one could construct a plant which would operate as a PMM 2. We see this by comparing Fig. 9.1 (c) with Fig. 8.3. We then note directly that the CHPP within control volume Z of Fig. 9.1 (c) is a Cyclic PMM 2, since no heat is rejected by it. Equally, as seen from Constrained System X, which executes a non-cyclic process, the plant within control volume Y is seen to constitute a Non-cyclic PMM 2, albeit containing a cyclic device within itself; this is because the plant within control volume Y has succeeded in taking the Constrained System (control volume X) from an initial Stable State to some other Allowed State as a result of the heat transfer Q to the CHPP while the sole net end effect external to X has been the delivery of positive work output, which could have been used to raise a weight.

Hence, it is impossible for the Effacer depicted in Fig. 9.1(b) to exist; according to the definition of an irreversible process given in Section 9.2, the original process depicted in Fig. 9.1(a) is therefore irreversible.

Having established by this routine procedure that an adiabatic throttling process is irreversible, we should stop to ask ourselves wherein lies the principal source of the irreversibility in this particular instance. It lies, in fact, in the random dissipation by eddies, *after* the orifice, of the kinetic energy that is generated as the fluid passes through the orifice. In this random dissipation of kinetic energy, the state of the fluid departs appreciably from equilibrium and then settles down to a state of near equilibrium as the fluid proceeds down the pipe towards station 2. As was noted in Section 2.14, it is the settling down from a non-equilibrium to an equilibrium state that constitutes the 'one-way' or 'irreversible' process. The flow up to the orifice is reasonably smooth or streamlined and so generates little irreversibility. This difference between the flow regimes upstream and downstream of the orifice is depicted in Fig. 9.1(a).

96

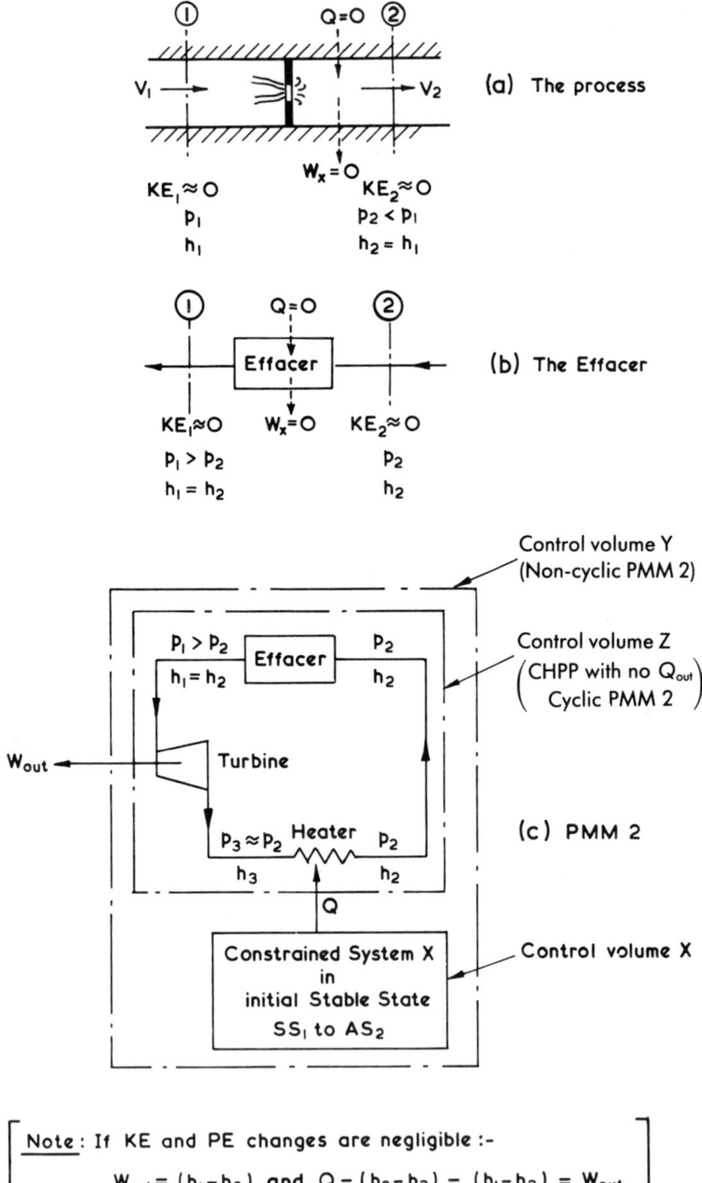

$$\left[\begin{array}{l}\text{Note: If KE and PE changes are negligible :-}\\ W_{out} = (h_1 - h_3) \text{ and } Q = (h_2 - h_3) = (h_1 - h_3) = W_{out}\end{array}\right]$$

Fig. 9.1 Demonstration that a steady-flow adiabatic throttling process is irreversible

9.5 Reversible processes

In Section 9.1, attention was drawn to the fact that all real physical processes are in some measure irreversible and can only approach full reversibility as a limiting ideal. From what has already been said, it will be recognized that to approach this ideal one must reduce to a minimum the unavoidable departures from equilibrium that inevitably occur in processes that have to be executed in a finite time. The ideal is only reached in 'Thermotopia', when processes are carried out infinitely slowly and the affected systems pass through a continuous succession of quasistatic Stable States. If one wishes to have a definition of reversibility, however, the following will meet the case as defining the opposite of irreversibility:

Definition

A process undergone by a system is *reversible* if all the effects of the process on the system and its environment can be effaced.

However, this definition, unlike that for an irreversible process, cannot be used as the basis for a routine test for reversibility. Although some texts claim to show that a given process is reversible, when they do so the authors are, in fact, simply using their acquired knowledge of what causes irreversibility to exclude all such possibilities from the process under demonstration. Thus we have a positive test for irreversibility but, if we are trying to ensure that a given process approaches as closely as possible to complete reversibility, we can only hope that we have recognized all possible sources of irreversibility and have then reduced their effects to the minimum that is feasible; we shall never eliminate their effects completely.

9.6 Principal categories of irreversible process

If we are to approach the ideal of full reversibility in our processes in order to obtain the best possible performance from our work-producing and work-absorbing devices, then we need to be able to recognize the principal sources of irreversibility; in other words, we need to be aware of the principal types of process which involve departures from equilibrium. To this end, by the procedure set out in Section 9.3 and illustrated in Section 9.4, we may show that the following processes come under this heading:

(a) *All frictional processes*
(b) *Unresisted expansion* (e.g. adiabatic throttling, or expansion of a gas behind a very fast-moving piston)
(c) *Heat transfer across a finite temperature difference*
(d) *The combustion of a fuel*
(e) *Diffusive mixing of fluid streams*, each at a different pressure or temperature, or each of different chemical composition.

The source of the irreversibility in process (b) above is basically no different from that in process (a), since the dissipation of the definitely-directed kinetic

98

energy by eddy formation and destruction after passage of the fluid through an orifice is the result of fluid friction. This itself is a reflection of the viscosity of the fluid. In a hypothetical non-viscous fluid, the jet emerging from the orifice would continue down the pipe indefinitely as a central core of fast-moving fluid. Viscosity also plays an essential part in diffusive mixing.

By the procedure set out in Section 9.3, the reader should be able to establish the irreversibility of the above processes, with the present exception of process (e), for the proof of which he will need to wait until the concept of a *semi-permeable membrane* has been introduced in Section 19.4.

9.7 Summary

In this chapter, as a preliminary to the further development of the theorems of equilibrium thermodynamics, we have discussed and defined two concepts of the greatest importance, namely the *irreversibility* of natural processes and the *reversibility* of ideal 'Thermotopian' processes. It is with the aid of the latter type of process that we shall later be able to set up criteria of excellence for our actual work-producing and work-absorbing devices.

Having defined an irreversible process as one whose effects on both the system in question and its environment cannot be effaced, we then devised a formal test for the irreversibility of a process. In this test, we supposed the process in question to be reversible and then examined whether, with the hypothetical effacing process (the *Effacer*) and other known natural processes, we could construct a PMM 2. If we succeeded in doing so, then we contravened Corollary 3 of the Law of Stable Equilibrium; we therefore concluded that the Effacer could not exist and the original process must consequently have been irreversible.

We pointed out that all real-life processes are in some measure irreversible because they have to be carried out in finite time and so cause the system to depart in some measure, however small, from a state of stable equilibrium (a Stable State). Since it might therefore seem that it was superfluous to devise a test for irreversibility, it was noted that we did this in order to shine a spotlight on those classes of process which are essentially irreversible; these were listed in Section 9.6. As we shall see in the next chapter, it is irreversible processes such as these that cause the performance of our real devices to fall below the ideal, frequently by a large amount.

Extension of the Thermodynamic Family Tree to cover the material discussed in this chapter is deferred to the end of the next chapter.

CHAPTER 10

Single-Reservoir Processes and Reversible-Work Theorems—Introduction to Thermodynamic Availability (With Appendix C)

10.1 Introduction

For want of a more scientific definition of temperature, we have so far been content with an arbitrary temperature, θ, such as might be recorded by a mercury-in-glass thermometer; this at least enables us to know when we have achieved constancy of temperature. In Sections 1.15.3, 6.5, and 8.9, attention was drawn to the fact that a scientific quantitative measure of the temperature of a system would have to await the definition of a property of a system which we call its *thermodynamic temperature* and to which we assign the symbol T. Before we can proceed directly to this task, we shall need to study the performance of reversible (and therefore hypothetical) cyclic devices exchanging heat with two *Thermal Reservoirs*. To this end, we find it convenient first to study the performance of devices which exchange heat with only a single Thermal Reservoir; for brevity, we shall call these *single-reservoir processes*. Such a study will not only prepare the ground for the development of the concept of thermodynamic temperature, but will also provide the starting point for a very important and virtually self-contained branch of equilibrium thermodynamics to which we give the title *thermodynamic availability*. We shall study that topic in detail in later chapters, where we shall regard the environment as akin to a single reservoir in the presence of which our work-producing and work-absorbing devices operate.

10.2 Thermal Reservoir

A *Thermal Reservoir* is a hypothetical device which nevertheless constitutes an indispensable item of the classical thermodynamicist's tool-kit. It is more usually given the name *heat reservoir*; this is a singularly unfortunate choice of words, since it conveys the false impression that heat resides in a body. For present purposes, we may define a Thermal Reservoir as follows:

100

Definition

A Constrained System which suffers only heat interactions with other systems, during which its temperature remains constant and it passes only through Stable States (so that all processes within it are reversible), is called a *Thermal Reservoir.*

Since the Constrained System in question suffers only heat interactions, a Thermal Reservoir comprises a system of fixed volume and of zero work input and output; it may therefore be depicted in the manner shown in Fig. 10.1(a) and (b), according to whether heat is transferred to or from the reservoir. That it is a purely hypothetical device follows from the stipulated requirement that it pass only through Stable States, since this implies that *all processes occurring within a Thermal Reservoir are reversible* and, as we saw in Section 9.1, such processes would occur only in 'Thermotopia'.

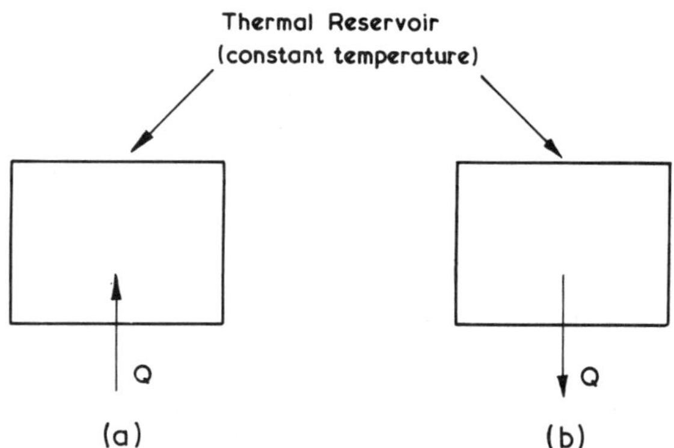

Fig. 10.1 Diagrammatic representation of a Thermal Reservoir

A Constrained System which was very large compared with other systems with which it exchanged heat at an infinitely slow rate could be considered as a near approximation to a Thermal Reservoir. Although the attainment of reversibility within such a system would be beyond us, being achievable only in 'Thermotopia', this fact need not worry us because a Thermal Reservoir is in any case simply a convenient hypothetical concept.

10.3 Datum Reservoir

We shall find it convenient to consider a Thermal Reservoir at one particular specified temperature as a *Datum Reservoir*; we shall give the symbol θ_d to its arbitrary temperature and shall later substitute for θ_d the corresponding thermodynamic temperature T_d.

10.4 First Reversible-Work Theorem – Gross work output in
non-cyclic, single-reservoir processes

As a first step towards developing the concept of thermodynamic temperature, we need to use our definition of a reversible process (Section 9.5) to prove an important theorem relating to the gross work produced by a system in a process between specified Stable States when the system can exchange heat with a single Thermal Reservoir (e.g. the Datum Reservoir at temperature θ_d). We shall later make a further important application of this theorem in Chapter 13 when we come to consider the topic of *thermodynamic availability*, which deals with the availability of energy for work production. We shall develop two theorems relating to gross work in these circumstances. We shall call this one the *First Reversible-Work Theorem*; it is stated as follows:

First Reversible-Work Theorem (*non-cyclic* processes)

For a system that can exchange heat with a single Thermal Reservoir (e.g. the Datum Reservoir at temperature θ_d), the gross work output is the same for all *fully reversible* processes between the same specified stable end states of the system. (This quantity is termed the *reversible gross work* and is given the symbol $[(W_g)_{\mathrm{REV}}]_1^2$ when the system is taken from Stable State 1 to Stable State 2.) During any irreversible process between these same specified end states during which the system may exchange heat with the same Thermal Reservoir, the gross work output is always less than $[(W_g)_{\mathrm{REV}}]_1^2$.

In passing, we may note that this proposition is equivalent to stating that, for a plant to produce maximum work output under specified conditions (or, *per contra*, to require minimum work input), all processes to which the plant is subject must be reversible. Proof of the proposition provides the theoretical basis for our statement in Sections 2.14 and 9.1 that *irreversibility always results in lost opportunities for producing work or in a greater work input than is ideally needed.*

We prove this Theorem by showing that, if it were not true, then we could construct a PMM 2, so contravening Corollary 3 of the Law of Stable Equilibrium that was proved in Section 8.2.

Figure 10.2(a) represents two alternative processes suffered by a system which is taken between specified Stable States 1 and 2, R being *fully reversible** and I irreversible. During these processes, the gross work outputs are respectively $(W_g)_R$ and $(W_g)_I$ while the system *rejects* heat quantities respectively equal to $(Q_d)_R$ and $(Q_d)_I$ to the Datum Reservoir. From the definition of a reversible process given in Section 9.5, since R is postulated to be reversible there must be a process E which could act as an effacer of all the effects of process R on both the system and its environment, which includes the Datum Reservoir; i.e. when the system was taken back to Stable State 1 from Stable State 2 in the effacing process E, the system

---

* The requirements for full reversibility are discussed in Section 10.6.

102

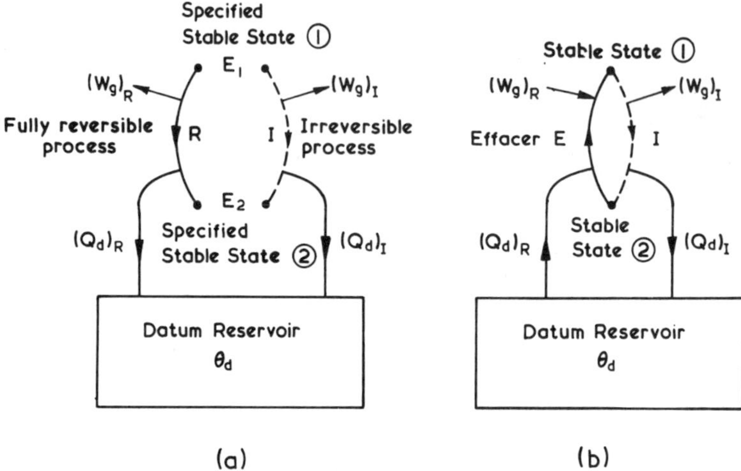

Fig. 10.2 Diagrams relating to the First Reversible-Work Theorem

would absorb from the environment work of magnitude equal to $(W_g)_R$ and would take in from the Datum Reservoir heat of magnitude equal to $(Q_d)_R$. Processes E and I together could then take the system round a completed cyclic process, as depicted in Fig. 10.2(b).

From the Energy Conservation Equation for the system (equation 7.6a),

$$[(Q_d)_R + (W_g)_R] = (E_1 - E_2) = [(Q_d)_I + (W_g)_I] . \tag{10.1}$$

Hence

$$[(W_g)_R - (W_g)_I] = [(Q_d)_I - (Q_d)_R] , \tag{10.2}$$

so that if $(W_g)_I$ were greater than $(W_g)_R$ then $(Q_d)_I$ would be numerically less than $(Q_d)_R$. In that event a net quantity of heat would be drawn from the Datum Reservoir in Fig. 10.2(b) and the plant would produce positive net work output in a completed cycle of the working fluid. Thus, recalling Fig. 8.3 and noting that the Datum Reservoir constitutes a Constrained System in an initial Stable State, we see that the cyclic process depicted in Fig. 10.2(b) would then be that of a Cyclic PMM 2 (while at the same time achieving precisely the effect required of a Non-cyclic PMM 2). This would contravene Corollary 3 of the Law of Stable Equilibrium, so that we must have

$$(W_g)_I \leqslant (W_g)_R . \tag{10.3}$$

However, if $(W_g)_I$ were to equal $(W_g)_R$ it would be possible to prove that process I was reversible, since process E could then also act as the effacer of process I. Hence

$$(W_g)_I < (W_g)_R . \tag{10.4}$$

Now, were I and R replaced in Fig. 10.2(a) by alternative fully reversible processes R_A and R_B respectively, expression (10.3) would become

$$(W_g)_{R_A} \leqslant (W_g)_{R_B},$$

whereas if they were replaced respectively by R_B and R_A it would become

$$(W_g)_{R_B} \leqslant (W_g)_{R_A}.$$

Since both of these statements are true, to satisfy them both we must have

$$(W_g)_{R_A} = (W_g)_{R_B}.$$

However, R_A and R_B are any two fully reversible processes between the specified end states, so that we may write

$$(W_g)_{R_A} = (W_g)_{R_B} \equiv [(W_g)_{REV}]_1^2. \tag{10.5}$$

Expressions (10.4) and (10.5) together establish the truth of the First Reversible-Work Theorem. The value of $[(W_g)_{REV}]_1^2$ will clearly be a function of the properties of the system in Stable States 1 and 2, and of the temperature θ_d of the Datum Reservoir. We shall develop this further in Chapter 13.

It is evident that for non-cyclic, single-reservoir processes of this kind between specified Stable States, the theorem will be equally applicable to finite and infinitesimal processes, a fact of which we shall take advantage in the proof of the Third Reversible-Work Theorem in Appendix C to this chapter.

10.5 Heat exchanged with the Datum Reservoir during fully reversible, *non-cyclic* processes between specified stable end states

Since the change in internal energy of the system of Section 10.4 is equal to $(E_2 - E_1)$ for all processes between the specified Stable States 1 and 2, it follows from equation (10.5) that

$$(Q_d)_{R_A} = (Q_d)_{R_B} \equiv [(Q_d)_{REV}]_1^2. \tag{10.6}$$

That is, *not only is the gross work output the same for all fully reversible processes between the specified Stable States 1 and 2, but so also is the heat quantity exchanged with the Datum Reservoir*. This important conclusion leads us on to the next step in our development of the concept of thermodynamic temperature.

In passing, we may also note from equation (10.2) that the *loss of gross work output due to irreversibility* is equal to the amount by which the heat rejected by the system to the Datum Reservoir is greater in the irreversible process than in the fully reversible process. This will give us a simple means of evaluating this loss of gross work output after we have devised a property called *entropy*, for which we first need to define *thermodynamic temperature*.

10.6 The requirements for full reversibility – Internal and external reversibility

The depiction of a fully reversible process in Fig. 10.2(a) is deceptively simple and calls for closer inspection. Recalling from Sections 6.8 and 9.6 that heat transfer across a finite temperature difference is irreversible, we conclude that full reversibility of process R of Fig. 10.2(a) can only be ensured when **either** (a) the system in question exchanges heat directly with the Datum Reservoir at temperature θ_d only when the system is itself at temperature θ_d (i.e. infinitesimally different from θ_d), as in Fig. 10.3(a), **or** (b) when heat exchange between a system at some temperature θ ($\neq \theta_d$) and the Datum Reservoir at temperature θ_d takes place via reversible auxiliary cyclic heat power plants (auxiliary CHPPs) where $\theta > \theta_d$, as depicted by the typical CHPP of Fig. 10.3(b). In the latter figure, the fluid circulating around the CHPP is at temperature θ when it takes in heat from the system at temperature θ and is at temperature θ_d when it rejects heat to the Datum Reservoir at temperature θ_d. (A CHPP is described as being reversible when all processes taking place within it are reversible. If θ were less than θ_d, we would need an auxiliary cyclic refrigerating plant instead of a CHPP.)

Thus *full reversibility* implies the existence of both *internal reversibility* within System Z and *external reversibility* of all heat exchanges between System Z and the Datum Reservoir by either the means depicted in (a) or those depicted in (b) of Fig. 10.3. Hence, we now see that the system which is subject to the fully reversible

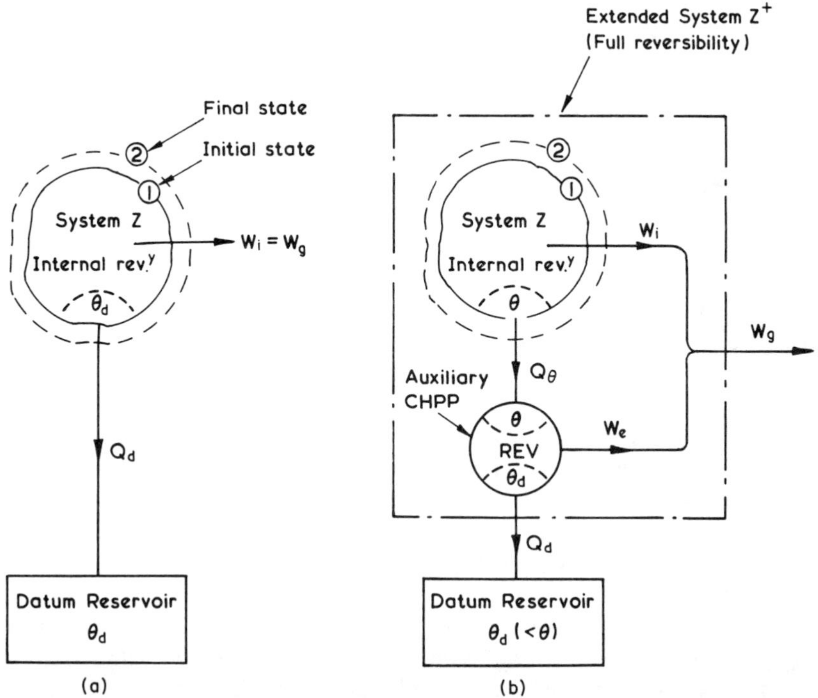

Fig. 10.3 Full reversibility = Internal reversibility + External reversibility

process in Fig. 10.2(a) would be System Z in Fig. 10.3(a) but the Extended System Z^+ in Fig. 10.3(b). Furthermore, in Fig. 10.3(a) the gross work output W_g would be simply the *internal work* W_i produced directly by System Z, whereas in Fig. 10.3(b) it would be the sum of W_i and the *external work* W_e produced by the auxiliary cyclic devices, of which that depicted in the figure is typical.

10.7 Internal work output in infinitesimal, non-cyclic processes

In Appendix C at the end of this chapter, there is derived a further theorem which we call the *Third Reversible-Work Theorem* and which relates to the work produced *directly* by a system (the *internal work*) in an *infinitesimal* process between specified stable end states. The reader should here take a look at this, if only superficially, since it highlights the difference between internal work output and gross work output. The theorem is put in an appendix because, in its proof and in the subsequent discussion, we use ideas which are only developed for the first time in later chapters. We shall need this Third Reversible-Work Theorem when, in Chapter 15, we come to evaluate the loss of work output due to irreversibility in the course of our study of the concepts and theorems of *thermodynamic availability*.

10.8 Second Reversible-Work Theorem — Fully reversible, *cyclic*, single-reservoir processes

In order to prove a further Theorem, which we shall find useful when considering the absolute zero of thermodynamic temperature in Section 11.5 and when introducing *entropy* in Chapter 12, we now consider the application of the First Reversible-Work Theorem to the special case in which the end state of the system is the same as the initial state, so that the system executes a fully reversible completed cyclic process while exchanging heat with a single Thermal Reservoir. The Theorem which we have to prove may be stated as follows:

Second Reversible-Work Theorem (*cyclic* processes)

For a system that executes a *fully reversible*, completed *cyclic* process while exchanging heat with a single Thermal Reservoir, the *net* output of gross work and the *net* heat quantity exchanged with the reservoir are both zero.

Since the process is fully reversible, the system must pass through a continuous succession of Stable States, so that, with reference to Fig. 10.4(a), we may conceptually 'break open' the cycle at some intermediate state 2 and imagine the system to be taken from the initial Stable State 1 to the intermediate Stable State 2, and then back to Stable State 1. We suppose that, in these two non-cyclic processes R_{12} and R_{21}, the gross work quantities produced and the heat quantities rejected by the system to the Datum Reservoir are respectively as shown in the figure. If some of these are positive, others will be negative.

Now, since process R_{12} is fully reversible, there will be a fully reversible effacing

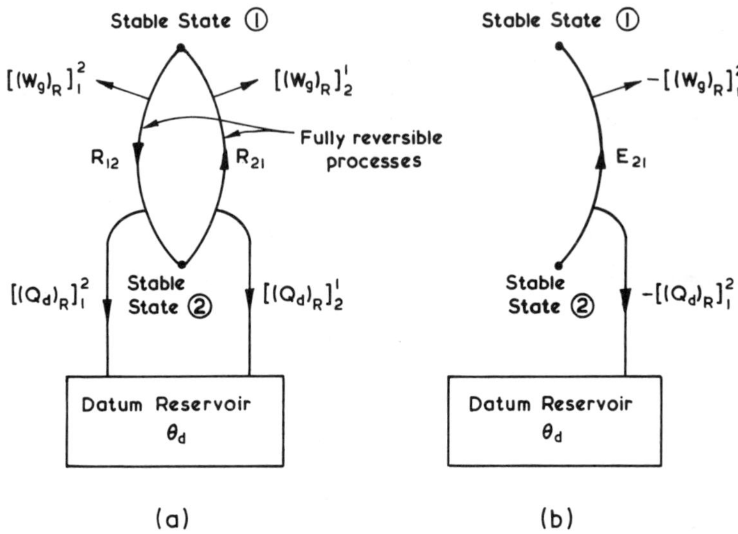

Fig. 10.4 Diagrams illustrating the proof of the Second Reversible-Work Theorem

process E_{21} during which, as indicated in Fig. 10.4(b), the system will deliver a quantity of work equal to $-[(W_g)_R]_1^2$ and reject a quantity of heat equal to $-[(Q_d)_R]_1^2$. But, from the First Reversible-Work Theorem, and in particular equations (10.5) and (10.6), these work and heat quantities must be equal to the corresponding work and heat quantities in the fully reversible process R_{21} of Fig. 10.4(a); that is,

$$- [(W_g)_R]_1^2 = [(W_g)_R]_2^1$$

and

$$-[(Q_d)_R]_1^2 = [(Q_d)_R]_2^1.$$

Hence,

$$[(W_g)_R]_1^2 + [(W_g)_R]_2^1 = 0 \tag{10.7}$$

and

$$[(Q_d)_R]_1^2 + [(Q_d)_R]_2^1 = 0. \tag{10.8}$$

Equations (10.7) and (10.8) together establish the truth of the above Theorem.

*10.9 Irreversible, cyclic, single-reservoir processes
and their relation to a Cyclic PMM 2

We shall appreciate more clearly the nature of a Cyclic PMM 2 (the hypothetical cyclic device whose impossibility of realization was proved in Section 8.6) if we spend a little time in examining the effect of replacing the fully reversible processes R_{12} and R_{21} in Fig. 10.4(a) by irreversible processes I_{12} and I_{21} respectively, as in Fig. 10.5. To avoid confusion and the use of negative signs, we shall here denote work and heat quantities *from* the system by W and Q respectively, while using W' and Q' for work and heat quantities *to* the system.

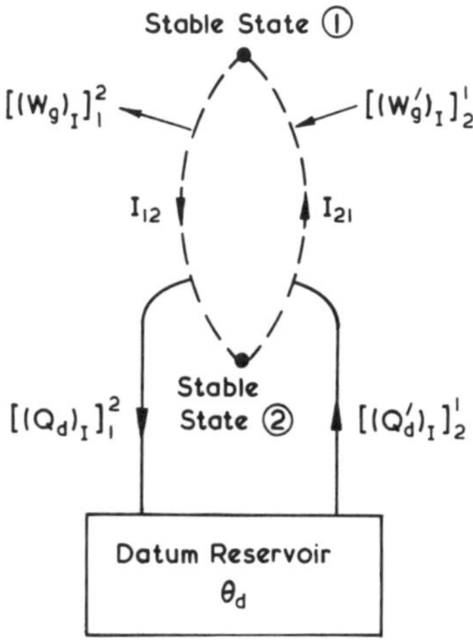

Fig. 10.5 Completed irreversible cyclic process with heat exchange with a single reservoir

From the First Reversible-Work Theorem, we know that $[(W_g)_I]_1^2 < [(W_g)_R]_1^2$ and that $[(W_g')_I]_2^1 > [(W_g')_R]_2^1$. Let us therefore define the *rational efficiency*† η_R of these respective irreversible processes in the following manner and, purely for purposes of convenient illustration, let us suppose that η_R has the same value for both, so that

$$[\eta_R]_1^2 \equiv \frac{[(W_g)_I]_1^2}{[(W_g)_R]_1^2} \qquad \text{and} \qquad [\eta_R]_2^1 = \frac{[(W_g')_R]_2^1}{[(W_g')_I]_2^1}. \tag{10.9}$$

The net *input* of gross work for the completed cyclic process comprising I_{12}

---

*This section could be ommitted at a first reading.

†We shall make further use of this term in our studies of *thermodynamic availability* in Chapter 13.

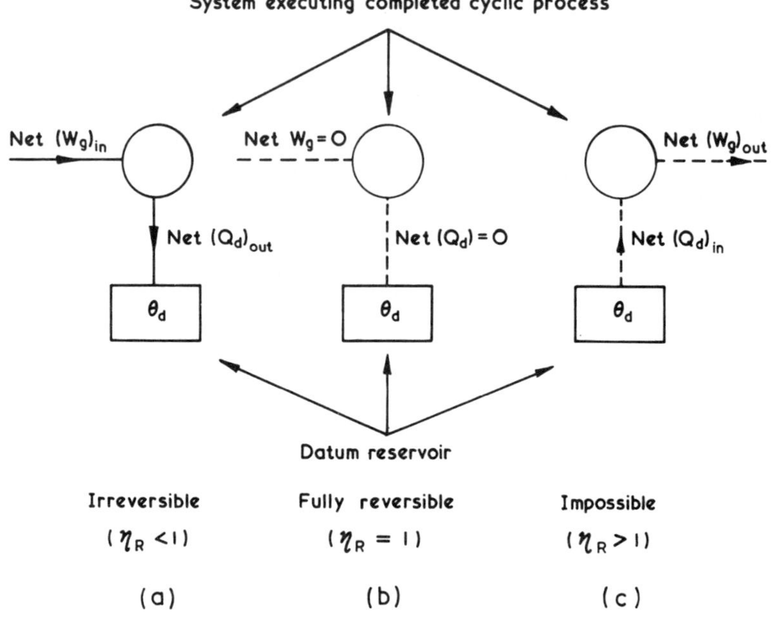

Fig. 10.6 Diagrammatic representation of a system executing a completed cyclic process while exchanging heat with a single Thermal Reservoir

followed by I_{21} is given by

$$\text{Net } (W'_g)_\text{I} = [(W'_g)_\text{I}]^1_2 - [(W'_g)_\text{I}]^2_1. \tag{10.10}$$

Since we know from the foregoing Theorem that $[(W'_g)_\text{R}]^1_2$ is numerically equal to $[(W'_g)_\text{R}]^2_1$, we have, from equations (10.9) and (10.10),

$$\frac{\text{Net } (W'_g)_\text{I}}{[(W'_g)_\text{R}]^2_1} = \left(\frac{1}{\eta_\text{R}} - \eta_\text{R}\right). \tag{10.11}$$

Since η_R is always less than unity, the net input of gross work in a completed cycle will always be positive. Hence a system which executes a completed cyclic process while exchanging heat with a single Thermal Reservoir will always operate in the manner depicted diagrammatically in Fig. 10.6(a). In 'Thermotopia', where the process would be fully reversible, Fig. 10.6(b) would represent its mode of operation. Since η_R can never exceed unity, operation in the mode depicted in Fig. 10.6(c) is impossible, a fact confirmed by the observation that Fig. 10.6(c) represents a PMM 2; from Corollary 3 of the Law of Stable Equilibrium, we know this to be impossible of realization (Sections 8.2 and 8.6).

An insight into just how close one could get to a PMM 2, without actually achieving such a feat, is indicated by Fig. 10.7, which was plotted from equation (10.11). The net input (or output) of gross work is just zero when $\eta_\text{R} = 1$, in con-

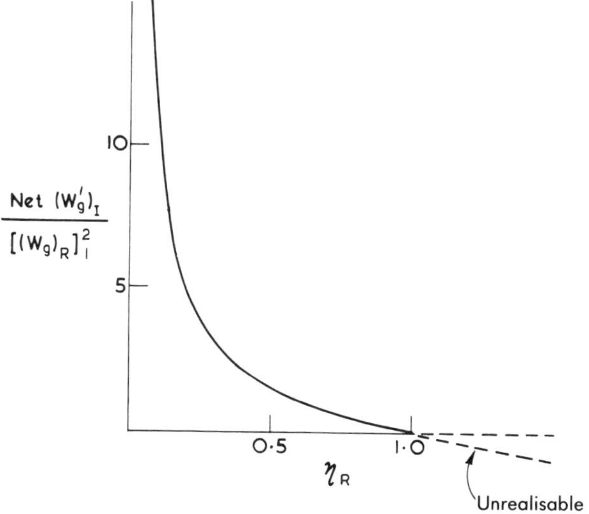

Fig. 10.7 Net $(W_g')_I/[(W_g)_R]^2_I$ plotted against η_R (equation 10.11)

formity with the above Theorem. Because η_R cannot exceed unity, we can never achieve a net output of gross work from a system which executes a completed cyclic process while exchanging heat with a *single* Thermal Reservoir; thus Fig. 10.7 demonstrates that we just fail to realize a PMM 2.

10.10 Summary

In this chapter we have shown in the *First Reversible-Work Theorem* that, when a system performs work while exchanging heat with a single *Thermal Reservoir* (Section 10.2), the gross amount of work produced when the system is taken from one specified Stable State to a second specified Stable State is the same for all ideal *fully reversible* processes between these two specified states. We have also shown that the gross work output is less than this unique value when the two states are joined by a process which does not meet the requirements for full reversibility set out in Section 10.6. To meet these requirements, it is necessary that there should be both *internal reversibility* within the system and *external reversibility*; the latter requires that the system should either exchange heat with the reservoir only when its temperature is infinitesimally different from that of the reservoir or that such heat exchange should take place via reversible auxiliary cyclic heat power (or refrigerating) plants. In the latter case, the *gross work* output is the sum of the *internal work* produced directly by the system and the *external work* produced by the auxiliary CHPPs. Since all natural processes are in some measure irreversible, it is this First Reversible-Work Theorem that enables us to reach the important conclusion that irreversibility always results in lost opportunities for producing work or in extra work input compared with the ideal. By using as the external source or sink

110

of heat the hypothetical ideal device which we have called a Thermal Reservoir, we avoid the introduction of any irreversibilities extraneous to the process under consideration since, by definition, a Thermal Reservoir undergoes only processes which are internally reversible.

This First Reversible-Work Theorem, which relates to *non-cyclic* processes between specified Stable States, provides the starting point for the development in Chapters 13 to 15 of the very important and virtually self-contained branch of classical thermodynamics known as *thermodynamic availability*. In the present chapter, however, we used it only to treat, in the *Second Reversible-Work Theorem*, the special case of a similar but *cyclic* process. In this, we showed that, when there is full reversibility in such a process, both the net output of gross work during a completed cycle and the net heat quantity exchanged with the reservoir are zero.

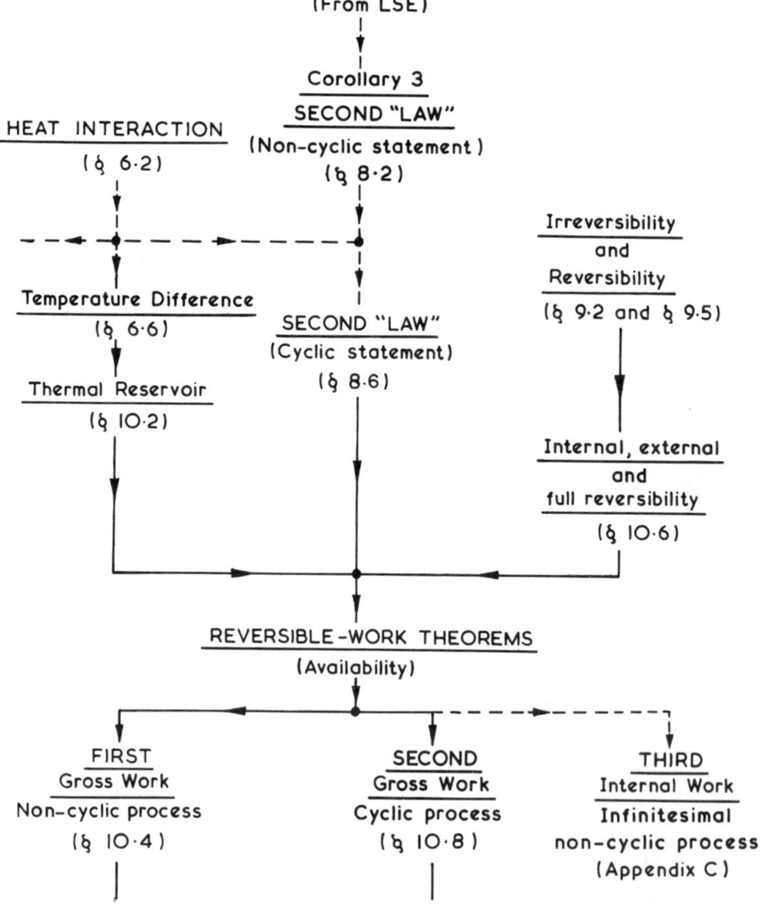

Fig. 10.8 Further development of the Thermodynamic Family Tree in Chapters 9 and 10 (Continuation from Fig. 8.5)

We shall find this theorem of great value when considering the *absolute zero* of thermodynamic temperature in Chapter 11 and again when introducing *entropy* in Chapter 12. In the present chapter we used it only as the basis for a discussion of the interesting question as to how closely it is possible to approach towards realization of the hypothetical devices which in Chapter 8 we described as Non-cyclic and Cyclic Perpetual Motion Machines of the Second Kind. A third theorem, the *Third Reversible-Work Theorem*, is developed and discussed in Appendix C at the end of this chapter.

The material covered in this and the previous chapter is depicted in the development of the Thermodynamic Family Tree shown in Fig. 10.8, which picks up from the point reached in Fig. 8.5.

APPENDIX C*
Third Reversible-Work Theorem

C.1 Internal work output in infinitesimal, non-cyclic processes

This Third Reversible-Work Theorem is presented here in an appendix for two reasons: first, because it follows on from the First Theorem and, second, because, in its proof and in the subsequent discussion, we have to use ideas which we shall not develop until we come to the next two chapters.

The First Reversible-Work Theorem of Section 10.4 related to the *gross* work output W_g from alternative processes between specified stable end states 1 and 2 of a system while the system exchanged heat with a Datum Reservoir at a constant specified temperature. In general, this temperature will be different from the temperature of the system, which will usually be variable. In proving that Theorem, we demonstrated that $[(W_g)_{REV}]_1^2$ is the same for all such fully reversible processes between these two Stable States. The same is not true for the work produced directly by the system, namely the *internal work* W_i, in that there is no unique quantity $[(W_i)_{REV}]_1^2$ which is the same for all *finite*, internally reversible processes between two specified states; the value of W_i in any such process in fact varies from process to process according to the path or nature of the process, as we shall shortly demonstrate. However, we shall now proceed to show that the internal work output $(\delta W_i)_{REV}$ is, *to a first order*, the same for all *infinitesimal*, internally reversible processes between two specified stable end states. Indeed, it is this fact that lies at the heart of the availability concept. We describe this proposition as the *Third Reversible-Work Theorem*, which we state formally as follows:

Third Reversible-Work Theorem (*Infinitesimal*, non-cyclic processes)

For a system that undergoes an *infinitesimal* change of state between specified Stable States, the work produced directly by the system (the *internal work*) is, *to a first order*, the same for all internally reversible processes between these two end states. (We give this quantity the symbol $(\delta W_i)_{REV}$.) During any internally irreversible process between these same specified end states, the internal work produced, $(\delta W_i)_I$, is always less than $(\delta W_i)_{REV}$.

We can establish the truth of this theorem as a corollary of the First Reversible-Work Theorem, which related to *gross* work output, W_g, in a process between

---

* This appendix could be omitted at a first reading.

specified Stable States. To do so, we suppose that, during any infinitesimal, internally reversible process R_A between the specified Stable States, System X of Fig. C.1 produces directly an output of *internal work* of magnitude $(\delta W_i)_{R_A}$ and that a heat quantity of magnitude $(\delta Q_i)_{R_A}$ is transferred from the system as its thermodynamic temperature changes from T to $T + \delta T$. For purposes of analysis, we then suppose that this heat transfer is made reversibly to a conceptual Thermal Reservoir, of temperature T, via a reversible CHPP which delivers a net *external work* quantity of $(\delta W_e)_{R_A}$, so resulting in a gross work quantity $(\delta W_g)_{R_A}$.

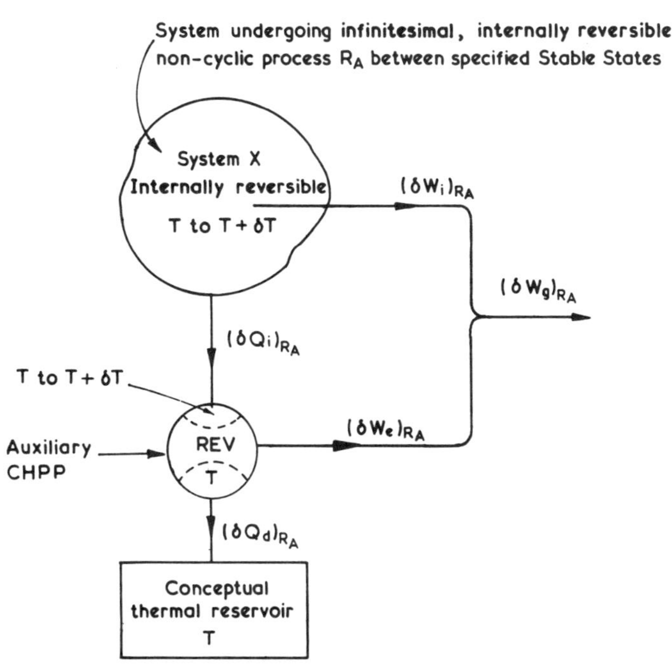

Fig. C.1 Diagram relating to the Third Reversible-Work Theorem (*Infinitesimal*, non-cyclic processes between specified Stable States)

Now, although the First Reversible-Work Theorem was derived for a finite process, it is clearly equally applicable to an infinitesimal process. If, therefore, the infinitesimal, internally reversible process R_A depicted in Fig. C.1 were replaced by an alternative internally reversible process R_B in similar circumstances, we know from the First Reversible-Work Theorem that we would have

$$(\delta W_g)_{R_A} = (\delta W_g)_{R_B} \equiv (\delta W_g)_{REV}. \qquad (C.1)$$

In the case of infinitesimal processes, however, the external work quantity is of second order. To establish this fact, we need to run ahead of ourselves and make use of the expression for the thermal efficiency of a Carnot cycle derived in Section

114

11.10 of the next chapter; for the reversible CHPP of Fig. C.1, we then see that

$$(\delta W_e)_{R_A} \simeq \frac{(\delta T/2)}{T + (\delta T/2)} (\delta Q_i)_{R_A}.$$ (C.2)

Since δT and δQ_i are both of first order, $(\delta W_e)_{R_A}$ is of second order. Hence we conclude that what is true precisely for δW_g is also true, but *to a first order*, for δW_i. Hence, noting expressions (10.5) and (10.4) relating to the First Reversible-Work Theorem, we may say that, to a first order,

$$(\delta W_i)_{R_A} \simeq (\delta W_i)_{R_B} \equiv (\delta W_i)_{REV}$$ (C.3)

and, for an infinitesimal, internally irreversible process between the same Stable States,

$$(\delta W_i)_I < (\delta W_i)_{REV}.$$ (C.4)

Expressions (C.3) and (C.4) together establish the truth of the Third Reversible-Work Theorem, which again leads us to the conclusion that irreversibility always results in lost opportunities for producing work or in extra work input compared with the ideal.

C.2 Discussion of infinitesimal, irreversible processes

There is a point of some subtlety which we have not yet touched upon in this study of infinitesimal processes between two Stable States. Since the processes are infinitesimal, we need to ask ourselves what is the essential factor that distinguishes an internally irreversible process from an internally reversible process in these circumstances. It would appear that the answer to this question must lie in the fact that, although the internally reversible process is infinitesimal, it will nevertheless need to be executed infinitely slowly if the process is to be truly reversible (or 'Thermotopian'), since we have already recognized in Section 9.1 that execution of any process in a finite time will inevitably lead to some departure from stable equilibrium and so to some measure of irreversibility, however small. All natural processes have necessarily to be completed within a finite time and so will always be irreversible, even when they bring about only an infinitesimal change of state from one Stable State to another. It is thus not a contradiction in terms to describe such an infinitesimal process as irreversible. Only in the ideal limit can a process be considered reversible.

In the foregoing discussion it must be borne in mind that it relates to internal equilibrium of a *finite* system on a macroscopic scale. While a very close approach to localized equilibrium at certain points within the system may well ante-date a similarly close approach to the complete establishment of a state of stable equilibrium throughout the system, it is execution of a process by a finite system within a finite time that causes disequilibrium between different parts of the system; this therefore results in irreversibility, for we have already noted that the settling down of an isolated system to a Stable State from a non-equilibrium state is an irreversible process. Thus, only if a process executed by a finite system is carried out infinitely slowly can a true state of stable equilibrium common to all parts of the system be maintained, so enabling the system to pass (in 'Thermotopia') through a continuous succession of Stable States. Classical thermodynamics does not concern itself with the shadowy borderline situation that arises when, in considering successively smaller finite systems, one tends ultimately towards the infinitesimal or ultramicroscopic in size.

It must also be recognized that, although we have introduced time into this discussion, time has not entered into any of our equations of equilibrium thermodynamics. The consideration of *rate processes* lies outside the field of classical equilibrium thermodynamics, but these form a subject of study in what is called *irreversible thermodynamics*, a topic which is beyond the scope of the present volume.

C.3 Discussion of the First and Third Reversible-Work Theorems

From these two theorems, it must be noted that it is only the gross work output that is the same for all *finite*, internally reversible processes between two specified Stable States. By contrast, whereas for *infinitesimal*, internally reversible processes

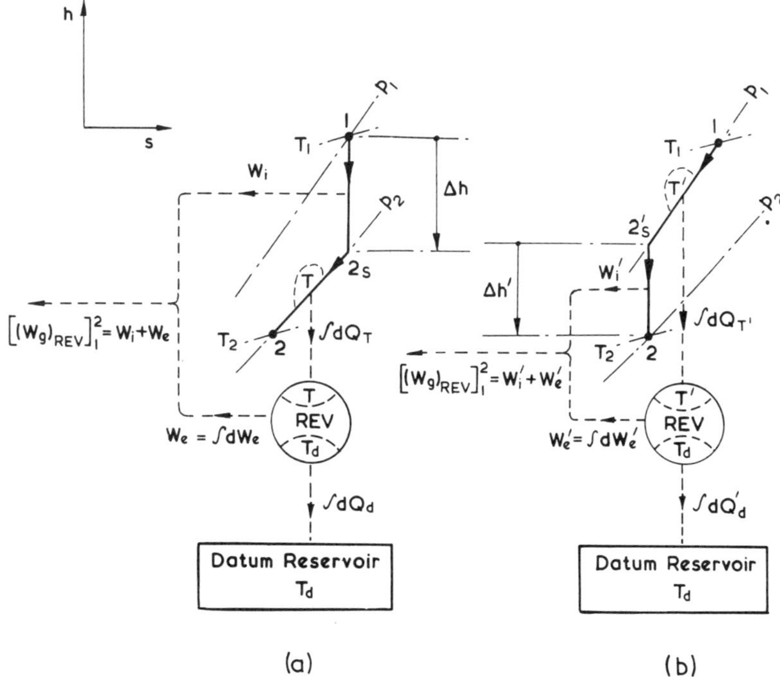

(a)   (b)

Fig. C.2 Diagrammatic representation, on an enthalpy–entropy diagram, of alternative finite, fully reversible processes between specified stable end states. (*Note:* From the second $T\,ds$ equation of Section 12.5 for a Simple System, namely $T\,ds = dh - v\,dp$, we have $(\partial h/\partial s)_p = T$ and $(\partial h/\partial p)_s = v$, so that isobars on the above enthalpy–entropy diagram steepen, and their vertical separation increases, as one moves to the right.)

the internal work output is, to a first order, the same for all such processes between two specified Stable States, this is not the case when the processes are finite. This contrast may be seen clearly in Fig. C.2, for an understanding of which one must have a knowledge of the enthalpy—entropy diagram for a Simple System (Section 5.3) presented in Section E.2 of Appendix E to Chapter 12.

Figure C.2 depicts, in highly diagrammatic form on an enthalpy—entropy diagram for a Simple System, two alternative finite, fully reversible, steady-flow processes between specified Stable States 1 and 2. The only internal work output is that produced in the isentropic process $1-2_s$ in (a) and $2_s'-2$ in (b), the magnitude of the internal work output $(W_i)_{REV}$ being in each case equal to the respective enthalpy drop when kinetic and potential energy changes are assumed to be negligible. It is immediately seen from these two enthalpy drops that $(W_i)_{REV}$ is not the same in (b) as in (a), for it will be evident from the footnote to Fig. C.2 that W_i' in (b) is less than W_i in (a). By contrast, however, the external work W_e' produced in (b) by the auxiliary cyclic devices needed to ensure external (and hence full) reversibility is greater than the corresponding quantity W_e in (a), since* $\delta W_e = (1 - T_d/T)\,\delta Q_T$, while $T' > T$ and $Q_T' > Q_T$. Thus, in changing from process (a) to process (b), the internal work output decreases but the external work output increases, so enabling the gross work output, which is the sum of these two, to be the same for both processes, in accordance with the First Reversible-Work Theorem.

We shall need to refer to the Third Reversible-Work Theorem when we come to evaluate the loss of work output (or extra work input) due to irreversibility; this we shall do in Chapter 15 in the course of our study of the concepts and theorems of thermodynamic availability.

* See Section 11.10.

CHAPTER 11

Two-Reservoir Processes and Thermodynamic Temperature

11.1 Introduction

Our next and final step towards the development of the concept of thermodynamic temperature is taken by applying to *two-reservoir processes* (viz. idealized cyclic heat power plants) the results for single-reservoir processes deduced in the last chapter. This we can do very easily by defining the chosen system by a boundary which includes within it both the cyclic heat power plant and one of the two Thermal Reservoirs.

Two-reservoir processes are of interest not only in this development of the abstract concept of thermodynamic temperature but also in real engineering plant. For example, a cyclic steam power plant bears a certain similarity to a two-reservoir process, for in it heat is taken in by the working fluid as it passes through the boiler and is rejected by the working fluid as it passes through the condenser. Such cyclic plant, both steam and gas, are treated in considerable detail in the author's earlier book.[10] Strictly speaking, however, they are not two-reservoir processes, since the heat sources and sinks are not Thermal Reservoirs in the sense defined in Section 10.2; nor are heat reception and rejection always at constant temperature.

11.2 Reversible CHPP exchanging heat with two Thermal Reservoirs

A cyclic heat power plant (CHPP) engaging in a two-reservoir process may be considered as a special case of the type of plant depicted in Fig. 10.3(b) of the last chapter, but in which, as depicted in Fig. 11.1, a second Thermal Reservoir of constant temperature θ takes the place of System Z of Fig. 10.3(b). The net work output W_{net} produced by the CHPP in a completed cycle is here equal to the gross work output W_g, since, by definition, a Thermal Reservoir neither produces nor absorbs any work.

Considering System Z^+, for a given value of Q_θ the upper Thermal Reservoir will always reach the same Stable State 2 *from a given initial Stable State 1*, since $Q_\theta = E_1 - E_2$. Hence, we know from the First Reversible-Work Theorem of Section 10.4 that the values of both W_{net} and Q_d will be the same for all fully reversible

117

118

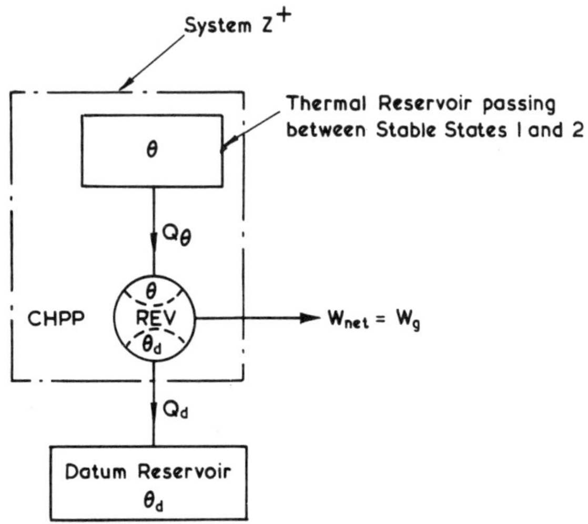

Fig. 11.1 Reversible CHPP operating between two Thermal Reservoirs

CHPPs in this given situation, *whatever their working substance*. Hence the ratio Q_θ/Q_d must be a function only of the temperatures θ and θ_d of the two reservoirs; that is,

$$\left(\frac{Q_\theta}{Q_d}\right)_{REV} = f(\theta, \theta_d). \tag{11.1}$$

11.3 Thermodynamic temperature function

There is an infinite variety of functions of θ and θ_d that would satisfy equation (11.1), but by international agreement one of the simplest possible has been chosen; in choosing this function the property called *thermodynamic temperature* is defined. The chosen function is

$$\left(\frac{Q_T}{Q_d}\right)_{REV} = \frac{T}{T_d}, \tag{11.2}$$

where T denotes thermodynamic temperature.

The choice of this particular functional relationship was not merely one of arbitrary convenience, for it had to be such as to be equally valid for the situation depicted in Fig. 11.2(a) as for that depicted in Fig. 11.2(b). By applying the First Reversible-Work Theorem (equations 10.5 and 10.6) to Systems A and B respectively, we know that $(W_1 + W_2) = W_3$ and $Q_d = Q_d'$ when, in both cases, the heat drawn from Thermal Reservoir 1 has the same value Q_1 and this reservoir starts from the same Stable State 1. From equation (11.2) we then have:

For CHPP 1,
$$\frac{Q_1}{Q_2} = \frac{T_1}{T_2}. \tag{11.3}$$

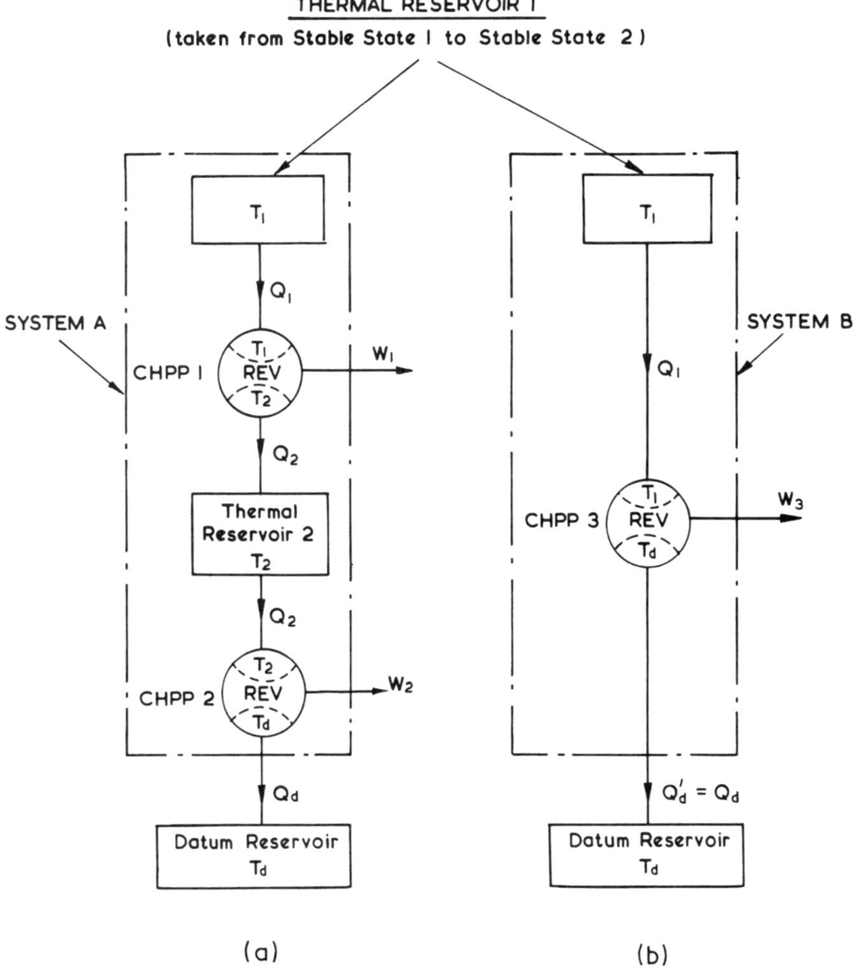

Fig. 11.2 Alternative situations for which the chosen thermodynamic temperature function must provide mutually acceptable solutions

For CHPP 2,
$$\frac{Q_2}{Q_d} = \frac{T_2}{T_d}.$$
(11.4)

For CHPP 3,
$$\frac{Q_1}{Q'_d} = \frac{Q_1}{Q_d} = \frac{T_1}{T_d}.$$
(11.5)

Now, from equations (11.3) and (11.4), we have
$$\frac{Q_1}{Q_d} = \frac{T_1}{T_d},$$
(11.6)

120

which agrees with equation (11.5). Thus our chosen thermodynamic temperature
function, namely equation (11.2), provides mutually acceptable solutions for both
situations (a) and (b) of Fig. 11.2. The reader will readily be able to devise a function which would not have done so in spite of having satisfied equation (11.1); it
would therefore not have been an acceptable function for defining thermodynamic
temperature.

11.4 A note on the use of the thermodynamic temperature function

It is clear from equation (11.2) that, if we choose an arbitrary Datum Reservoir
of accurately reproducible temperature level and assign an arbitrary numerical value
to its thermodynamic temperature T_d, we may, *in principle*, determine the value
of the thermodynamic temperature T of any other Thermal Reservoir by operating
a fully reversible CHPP between the two reservoirs and measuring Q_T and Q_d over a
completed cycle. We can only do this in principle because we cannot build fully
reversible cyclic heat power plants. We shall later devise, in Chapter 18, methods of
overcoming this limitation by the use of theoretical thermodynamic relations, but
it will still remain true that the only temperature level at which we shall have an
exact numerical value for the thermodynamic temperature will be that of our
chosen Datum Reservoir, to which we arbitrarily assigned the exact value T_d; the
thermodynamic temperature at all other temperature levels can never be known
exactly and any experimentally determined values will thus have to be quoted as
subject to an estimated uncertainty or tolerance.

Because of this situation, it has been necessary to set up, by international agreement, an International Practical Temperature *Scale*, which will be discussed briefly
in Sections 11.8 and 11.9. It should be particularly noted, therefore, that true
thermodynamic temperatures are defined by a *function*, namely equation (11.2),
and *not* by a scale, though, by long habit, many texts unfortunately quote equation
(11.2) as defining a thermodynamic temperature 'scale'. Attention was first drawn
to the erroneous nature of that practice in the author's paper of 1967 quoted under
Ref. 30 in the **ADDENDA TO REFERENCES**.

11.5 The Absolute Zero of thermodynamic temperature

Figure 11.3 depicts the means by which, in principle, the thermodynamic temperature T of a reservoir could be determined from measurements of Q_T and Q_d
when respectively $T > T_d$, as in (a), and $T < T_d$, as in (b).

Concentrating our attention on (b), we see from equation (11.2) that the value
of T will be given by

$$T = \frac{Q_T}{Q_d} T_d. \tag{11.7}$$

Hence, for a given value of Q_d, the value of Q_T will be smaller the lower is the value
of T, and as we take reservoirs of lower and lower temperature, so Q_T will tend
towards zero. The reservoir of zero thermodynamic temperature will thus be one
for which Q_T just reaches the vanishing point. At first sight it would appear that
when it does so the CHPP would become a Cyclic PMM 2; however, from the

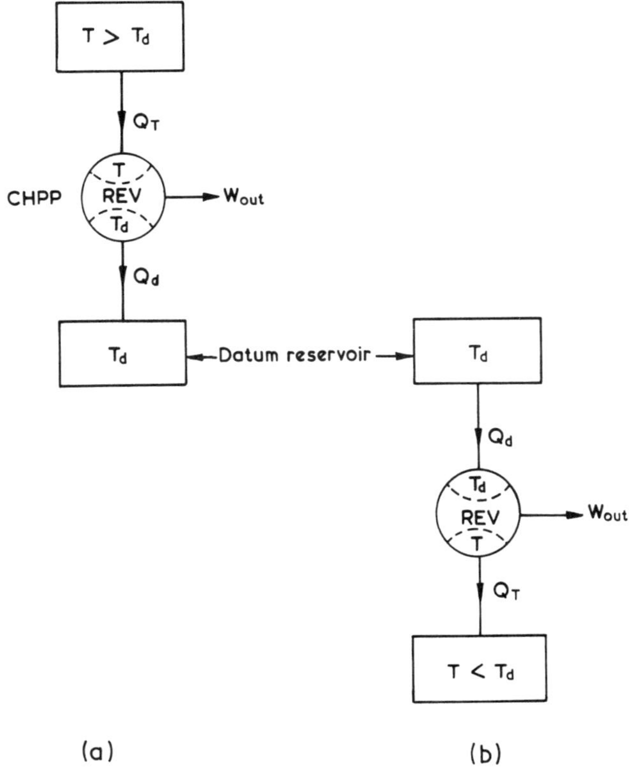

Fig. 11.3 Operation of CHPPs between the Datum Reservoir and reservoirs for which respectively $T > T_d$ and $T < T_d$

Second Reversible-Work Theorem of Section 10.8 we know that, since the CHPP is fully reversible, we would then also have

$$Q_d = W_{out} = 0,$$

so that, in effect, the CHPP would itself also vanish.

The fact that, when $T = 0$, $Q_T = 0$ and we then appear to have a Cyclic PMM 2 (a hypothetical device which, from Corollary 3 of the Law of Stable Equilibrium and the deduction therefrom in Section 8.6, we know cannot exist) is frequently quoted as an indication that it is not possible to reach absolute zero of thermodynamic temperature. From the foregoing discussion, however, we see that we could apparently *just* reach absolute zero without having to call in aid a Cyclic PMM 2. We therefore have to find surer grounds for asserting the unattainability of absolute zero. Before doing so, however, we might ask ourselves whether, in the first place, there is any good reason for expecting that we should be able to reach absolute zero. If we go back to the reasoning from which we set up the thermodynamic temperature function, namely equation (11.2), we see that we could just

122

as well have defined thermodynamic temperature in terms of the function

$$\left(\frac{Q_\tau}{Q_d}\right)_{\text{REV}} = \frac{\tau_d}{\tau}, \tag{11.8}$$

where $\tau \equiv 1/T$. This was indeed considered as a possibility by Lord Kelvin, the originator of the concept of thermodynamic temperature; in fact, it appears as a more natural way of expressing thermodynamic temperature when approaching the problem in terms of statistical thermodynamics. Now zero value of T represents an infinite value of τ. Since we have no expectation of ever attaining the latter, we ought not to be surprised to find that we can have no expectation of ever being able to reach zero T. Let us, therefore, finally approach the question from a more practical point of view.

Instead of the Thermal Reservoir of constant temperature T in Fig. 11.3(b), let us consider a Constrained System whose temperature we proceed to lower by extracting heat continuously from it by the Cyclic Refrigerating Plant depicted in Fig. 11.4, which discharges heat to the Datum Reservoir of constant temperature T_d (at, say, about atmospheric temperature). Since $dW_{\text{in}} = dQ_d - dQ_T$, we have, from equation (11.7),

$$dW_{\text{in}} = \left(\frac{T_d}{T} - 1\right) dQ_T. \tag{11.9}$$

Hence, as T tends towards absolute zero, the required work input to the Cyclic Refrigerating Plant per unit of heat drawn from the low-temperature reservoir tends towards infinity, even for this 'Thermotopian' fully reversible plant. We thus see that while we can approach absolute zero within apparently very close limits (perhaps, by appropriate means, within one-thousandth of a degree or closer), it will always represent a limiting value beyond our ultimate reach.* Even if we were to suppose that we had been able, by some means unknown, to bring the Constrained System down to absolute zero, then in order to maintain it there we would require an infinite amount of work to extract even the minutest amount of heat that had leaked into it from the environment. Absolute zero nevertheless constitutes a definite, fixed level of temperature. Having established this fact, we are now in a position to define the unit of thermodynamic temperature.

---

* This fact has been expressed in terms of the following proposition, sometimes described as the *Third Law of Thermodynamics*:

> *It is impossible by any procedure, no matter how idealized, to reduce any system to the absolute zero of temperature in a finite number of operations.*

It can be shown[11] that this statement is equivalent to the statement of the *Nernst Theorem* (or *Third 'Law'*) presented in Section 20.10.

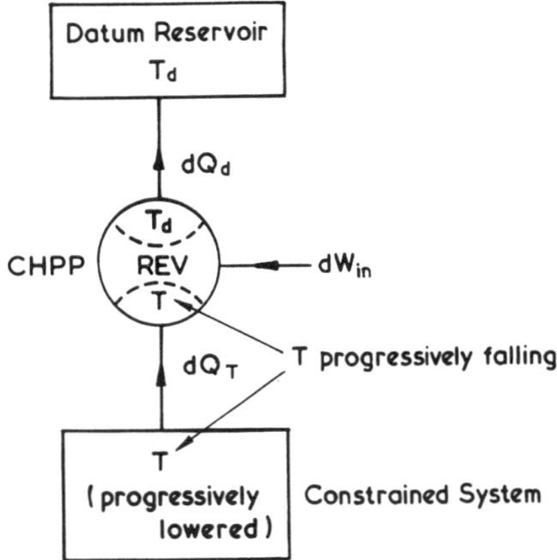

Fig. 11.4 Progressive reduction in temperature of a Constrained System by means of a reversible Cyclic Refrigerating Plant

11.6 The unit of thermodynamic temperature

Having chosen our thermodynamic temperature function, equation (11.2), and having established the existence of an absolute zero of thermodynamic temperature, our next step is to choose a convenient Datum Reservoir and then to assign an arbitrary (though also convenient) number of units, which we call *kelvins*, to this particular temperature level. In this way we shall define the magnitude of the *unit* of thermodynamic temperature.

As our Datum Reservoir, we must choose a physical set-up which can be relied upon to give us a convenient and accurately reproducible temperature level; for this purpose we choose that of a water *triple-point cell*, containing an equilibrium mixture of the solid, liquid, and vapour phases of water substance at the triple point (Section A.4). We find this convenient because the temperature of water at its triple point is only about one-hundredth of a Celsius degree (which we shall equate to a kelvin) higher than that of the *ice point* (the freezing point of air-saturated water at a pressure of 1 atm), and this was historically the arbitrary zero point of the so-called Celsius (or Centigrade) scale. The latter zero point has now passed into history, but we assign to the thermodynamic temperature T_d of our chosen Datum Reservoir such a value as to maintain close continuity of numerical values between the old and the new; to this end we assign a numerical value of *exactly* 273.16 kelvins to the triple-point temperature of water.

The foregoing procedure in effect defines the unit of thermodynamic temperature, to which we give the name *kelvin* and the unit symbol K (not °K, which is

124

now obsolete). Its formal definition is as follows:

Definition

The unit of thermodynamic temperature called the *kelvin* is the fraction **1/273.16** of the thermodynamic temperature of the triple point of water. (Defined by the Thirteenth General Conference of Weights and Measures, 1967.)

11.7 Celsius temperature

Scientifically, it would be a great convenience if temperatures were never expressed in anything but kelvins, but, in order to satisfy the practical demands of long-established habit, it is often convenient to quote temperatures in the form of a truncated thermodynamic temperature called the *Celsius temperature*, defined by the expression

$$t = T - 273.15 \; exactly, \tag{11.10}$$

where $t\,°C$ is the Celsius temperature corresponding to the thermodynamic temperature of $T\,K$. This can lead to a certain confusion in nomenclature, so that the following points should be noted:

(a) The sign $°C$ is not a unit symbol of the *Système International* (as is the symbol K for the unit of thermodynamic temperature called the kelvin).
(b) The quantity $t\,°C$ is often read as 't degrees Celsius', but 't Celsius' is to be preferred. (Were the term 'degree Celsius' to be thought of as a unit of temperature difference, then it is evident that, in this respect, it would be synonymous with the kelvin.)
(c) A temperature *difference* is always expressed in *kelvins*; for example, $30\,°C - 20\,°C = 10\,K$.

It should also be noted that the following terms and symbols are now obsolete:

Thermodynamic temperature *scale*

Centigrade

$°K$, degK, degC

11.8 The need for an International Practical Temperature Scale

We saw in Section 11.4 that, in order to determine the value of the thermodynamic temperature T of any Thermal Reservoir in general, one would *in principle* be able to do so by operating a fully reversible CHPP between the reservoir in question and the Datum Reservoir at $T_d = 273.16\,K$, measuring Q_T and Q_d over a completed cycle, and thence calculating T from the defining function, equation (11.2). However, since a fully reversible CHPP is a 'Thermotopian' device not capable of practical realization, the only temperature level at which the *exact* number of kelvins is known is that in terms of which the kelvin is defined, namely the triple point of water. Consequently, at all other temperature levels only *best estimates* of the number of kelvins can be made (by a combined application of theory and experiment, as will be discussed in Chapter 18). For practical purposes, it is therefore necessary to devise a *practical temperature scale* in which, at each of a number of accurately reproducible temperature levels (called *fixed points*), there is *assigned*, by international agreement, a specified number of kelvins, and in which methods of interpolation on the scale are also specified. The number of kelvins assigned at each fixed point is that which, by international agreement at the time, was considered to be the best estimate of the true thermodynamic temperature at the specified temperature level. The last such assignment was made in 1968 and superseded that previously made in 1948/1960. An amended edition of the 1968 Scale was issued in 1975, but this only introduced minor amendments which did not involve any changes in temperatures measured on the 1968 Scale.

11.9 The International Practical Temperature Scale of 1968 (IPTS-68)

The IPTS-68[12] *is the temperature scale in every-day use.* It distinguishes between the *International Practical Kelvin Temperature*, with the symbol T_{68}, and the *International Practical Celsius Temperature*, with the symbol t_{68}; similarly to equation (11.10), these are such that

$$T_{68} - t_{68} = 273.15 \text{ K } exactly. \qquad (11.11)$$

Thus for both of these the unit of temperature is the kelvin.

When expressing International Practical Celsius Temperature, use is again made of the sign $^{\circ}$C, so that, for example, an International Practical Celsius Temperature of 100 $^{\circ}$C corresponds to an International Practical Kelvin Temperature of 373.15 K. The difference between $t_a\,^{\circ}$C and $t_b\,^{\circ}$C is $(t_a - t_b)$ kelvins, namely $(t_a - t_b)$ K.

Two of the *defining fixed points* on the Scale and their *assigned* values are:

$$\text{Triple point of water} \qquad = 273.16 \text{ K} = 0.01\ ^{\circ}\text{C},$$

$$\text{Boiling point of water at 1 atm} = 373.15 \text{ K} = 100\ ^{\circ}\text{C}.$$

Complete specification of the Scale includes the statement of precise assigned values at other defining fixed points ranging from 13.81 K ($- 259.34\ ^{\circ}$C) at the triple point of equilibrium hydrogen to 1337.58 K (1064.43 $^{\circ}$C) at the freezing point of gold. The Scale defines further *secondary reference points* (some of whose values were changed in the 1975 edition[13] of IPTS-68) and these carry the Scale up to 3695 K (3422 $^{\circ}$C) at the temperature of melting tungsten. The *ice point* (the temperature of equilibrium between ice and *air-saturated* water) is one of these secondary reference points and has the assigned value of 273.15 K (0 $^{\circ}$C).

With the exception of the assigned value at the triple point of water (which is exact, by definition of the kelvin), the assigned values at all other fixed points correspond only to the best current estimate of the true thermodynamic temperature, in kelvins, at each fixed point. Estimated uncertainties of the assigned values

126

at the defining fixed points were quoted in the 1968 edition of the Scale, but these were withdrawn in the 1975 edition because more recent work had indicated that some of the uncertainties had been underestimated. In practice, we ignore these uncertainties.

11.10 The Carnot Cycle – Thermal efficiency

A hypothetical CHPP operating on the *Carnot Cycle* is one in which all processes are internally reversible, the working fluid receives heat (say Q_1) only at one constant temperature (say T_1) and rejects heat (say Q_2) only at a second lower constant temperature (say T_2). From the discussion in Section 11.2 and the definition of thermodynamic temperature T in Section 11.3, we thus have for the Carnot cycle:

$$\frac{Q_1}{Q_2} = \frac{T_1}{T_2}.$$

Now the net work W_{net} produced in the cycle is given by

$$W_{net} = Q_1 - Q_2,$$

and if we define the *thermal efficiency* η_{Carnot} of the cycle as

$$\eta_{Carnot} \equiv \frac{W_{net}}{Q_1},$$

we then have

$$\eta_{Carnot} = \left(1 - \frac{Q_2}{Q_1}\right) = \left(1 - \frac{T_2}{T_1}\right). \tag{11.12}$$

This expression for the thermal efficiency of the Carnot cycle is independent of the nature of the working fluid and follows from the manner in which thermodynamic temperature is defined.

11.11 The Carnot efficiency as a limiting value

With the aid of the First Reversible-Work Theorem of Section 10.4, we may readily show that the thermal efficiency of an irreversible CHPP* operating between two Thermal Reservoirs of specified temperature will necessarily be less than that of a reversible CHPP operating between the same two reservoirs; i.e. its thermal efficiency will be less than the Carnot efficiency for the specified reservoir temperatures.

Let us suppose that, as depicted in Fig. 11.5, both plants draw the same quantity of heat Q_1 from the high-temperature reservoir, which starts from the same

---

* When we say that a CHPP is irreversible, we mean that the *process* through which it passes is irreversible.

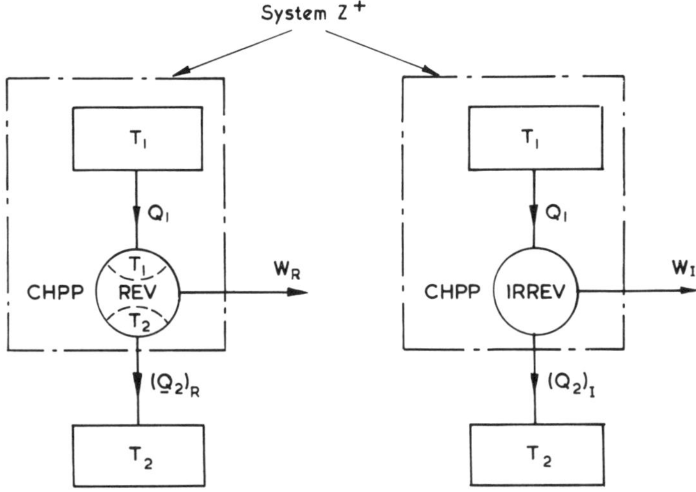

Fig. 11.5 The Carnot efficiency as a limiting value

initial Stable State in both cases. Then System Z^+, while exchanging heat with a single reservoir (that at T_2), will be taken through the same change of state in the two cases, there being no change of state of the CHPPs during an integral number of complete cycles. The First Reversible-Work Theorem then tells us that $W_I < W_R$. Hence, since $\eta_R = \eta_{\text{Carnot}} = W_R/Q_1$ and $\eta_I = W_I/Q_1$, we shall have

$$\eta_I < \eta_{\text{Carnot}}. \tag{11.13}$$

11.12 Cyclic refrigerators and heat pumps

A work-producing CHPP operating between two Thermal Reservoirs draws heat from a high-temperature reservoir, rejects a smaller quantity of heat to a reservoir of lower temperature and produces a net work *output*. By suitable choice of the working fluid passing round the cycle, we may alternatively arrange for a cyclic device, while absorbing a net work *input*, to draw heat from a low-temperature reservoir and deliver a larger quantity of heat to a reservoir of higher temperature. We then have a cyclic refrigerating plant. In both cases, the heat transfer to the fluid circulating round the cyclic device of course occurs while the fluid is at a lower temperature than the reservoir from which it is drawing heat; likewise, in both cases, heat rejection from the circulating fluid occurs when it is at a higher temperature than the reservoir to which it rejects heat. Such a cyclic refrigerating plant is depicted in Fig. 11.6.

When interest centres on the extraction of heat from the lower-temperature reservoir A (e.g. when A represents a refrigerating chamber and B the atmosphere), the plant is called a *refrigerator*. On the other hand, when interest centres on the supply of heat to the higher-temperature reservoir B (e.g. when B represents a room to be heated and A the atmosphere or a supply of river water at a temperature *lower*

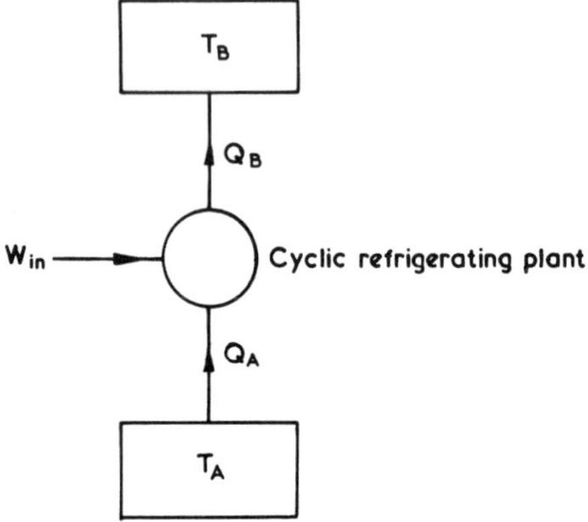

Fig. 11.6 Two-reservoir cyclic refrigerating plant (refrigerator or heat pump)

than that of the room), it is customary to describe the plant somewhat misleadingly as a *heat pump*.

In place of the thermal efficiency of a CHPP, we now have a *coefficient of performance* CP, defined respectively as follows for the two types of plant:

$$\text{Refrigerator CP} \equiv \frac{Q_A}{W_{in}}, \tag{11.14}$$

$$\text{Heat-pump CP} \equiv \frac{Q_B}{W_{in}}. \tag{11.15}$$

For the heat pump, the name *performance energy ratio* is sometimes used in place of coefficient of performance.

When all heat taken in by the working fluid in such refrigerating plant is taken in at a constant temperature T_A and all heat is rejected at a constant temperature T_B, and the cycle is internally reversible, the plant operates on what may be described as a *reversed Carnot cycle*. From the foregoing definitions, the coefficients of performance will then be respectively as follows:

$$\text{Refrigerator CP} = \frac{T_A}{T_B - T_A}, \tag{11.16}$$

$$\text{Heat-pump CP} = \frac{T_B}{T_B - T_A}. \tag{11.17}$$

11.13 Summary

In this chapter, as our final step towards the development of the concept of thermodynamic temperature as a property of a system, we used the single-reservoir

First Reversible-Work Theorem of Section 10.4 to arrive at certain important conclusions relating to the behaviour of ideal, fully reversible cyclic heat power plants exchanging heat with two Thermal Reservoirs; this we did by defining a boundary which contained within it both the CHPP and one of the two reservoirs. By this means, we showed that the ratio of the heat quantities exchanged by the ideal CHPP with two reservoirs of different temperature is the same for all fully reversible CHPPs operating between these two reservoirs, regardless of the nature of the working substance of the CHPP. This ratio must therefore be a function only of the temperatures of the two reservoirs; in accordance with internationally accepted practice, we chose the appropriately simple *thermodynamic temperature function*

$$\frac{Q_T}{Q_d} = \frac{T}{T_d},$$

where the subscript d relates to an arbitrary Datum Reservoir.

With the aid of the Second Reversible-Work Theorem of Section 10.8 we established the existence of an *absolute zero* of thermodynamic temperature. This enabled us to define the SI unit of thermodynamic temperature, called the *kelvin*, by choosing as the temperature of our Datum Reservoir that of a water triple-point cell and assigning to it the exact value of 273.16 kelvins (273.16 K). We then defined the truncated thermodynamic temperature known as the *Celsius temperature* through the relation

$$t = T - 273.15 \text{ exactly.}$$

Finally, we saw that, because the thermodynamic temperature function is defined in relation to the performance of ideal, fully reversible CHPPs and we are not able to build these, the only temperature level at which the exact number of kelvins is known is that in terms of which the kelvin is defined, namely the triple point of water. Thus, it was pointed out that, for practical purposes, it was necessary to devise an *International Practical Temperature Scale* in which the number of kelvins assigned at each *fixed point* was that which, by international agreement, was considered to be the best current estimate of the true thermodynamic temperature at that fixed point. The method by which such best estimates can be made will be discussed briefly in Chapter 18.

The chapter closed with a brief discussion of hypothetical, reversible, two-reservoir power and refrigerating plants operating respectively on the Carnot and reversed-Carnot cycles, the thermal efficiency of the former and the coefficient of performance of the latter being expressed in terms of the thermodynamic temperatures of the two reservoirs.

Extension of the Thermodynamic Family Tree to cover the material discussed in this chapter is deferred to the end of the next chapter.

CHAPTER 12

Entropy
(With Appendix D and Appendix E)

12.1 Introduction

Having defined thermodynamic temperature, we are now in a position to use the
First and Second Reversible-Work Theorems of Sections 10.4 and 10.8 to introduce
the concept of *entropy*, a very important thermodynamic property of a system
which is also of great utility. As introduced in classical thermodynamics, entropy
unfortunately appears as a very abstract concept and its basic nature can only be
properly understood in terms of the propositions of statistical thermodynamics.
For that reason its physical significance is not easy to grasp, but practice in its use
and application will soon breed familiarity and competence, if not full understanding. We shall find that the following are amongst the valuable features of
entropy that make it an indispensable tool to the engineer:

(a) If the path of an internally reversible process is sketched on a diagram with
    entropy as abscissa and thermodynamic temperature as ordinate, the area
    below the path is a measure of the heat exchanged by the system during the
    process.
(b) In a reversible, adiabatic process the entropy of the system remains
    constant.
(c) As we noted in the last paragraph of Section 10.5, entropy provides us with
    a very simple means of evaluating the loss of work output (or extra work
    input) due to irreversibility. By this means we are readily able to quantify
    the loss of performance resulting from irreversibilities in our work-produc-
    ing and work-absorbing plants.

12.2 The derivation of entropy as a thermodynamic
property of a system

As was noted in Section 1.14, it follows from the definition of a thermodynamic
property that the change in value of any property P during any process undergone
by a system between specified end states is dependent only on those end states; i.e.
it is independent of the nature or path of any process that carries the system
between the specified end states, so that this change in value is the same for all
processes by which the given change of state can be brought about. Bearing this

fact in mind, a clue to the discovery of entropy as a thermodynamic property is provided in the First Reversible-Work Theorem of Section 10.4. There we saw that, for fully reversible processes in the situation to which that theorem relates, the gross work is the same for all such processes between the same specified end states. In Section 10.5 we went on to show that the heat exchanged with the Datum Reservoir in these circumstances is also the same. Accordingly, we need to set up a situation in which we can make use of this fact.

Let System Z in Fig. 12.1(a) be the system for which we have to discover the property called entropy. We suppose it to undergo an infinitesimal, internally reversible process during which *internal work* $(dW_i)_{REV}$ (as defined in Section 10.6) is produced directly by the system while the latter absorbs heat $(dQ_T)_{REV}$ when, at the instant shown, it is at a thermodynamic temperature T.

In order to set up a situation to which we can apply the First Reversible-Work Theorem, we further suppose that the heat taken in by System Z is supplied from a reversible cyclic heat power plant which in turn absorbs heat $(dQ_d)_{REV}$ from an arbitrary Datum Reservoir at constant temperature T_d, as depicted in Fig. 12.1(b). The Extended System Z^+ then undergoes a fully reversible process, just as did the corresponding system in Fig. 10.3(b). The choice of T_d is purely arbitrary and, for convenience, we have taken it to be of such a value that it is greater than T at all times. We could equally well have taken T_d to be less than T and we would then have needed a reversible refrigerating plant (or 'heat pump') in place of the CHPP.

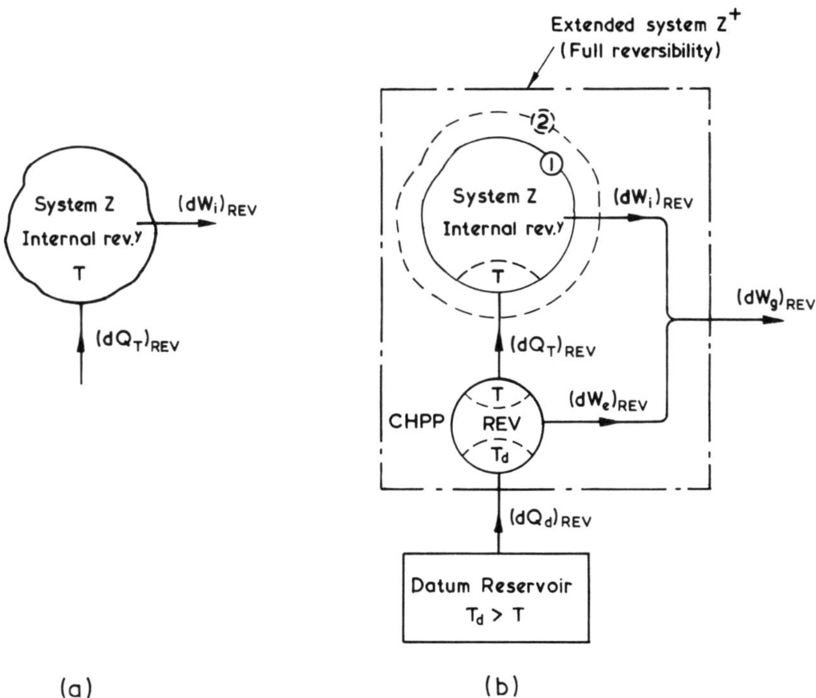

Fig. 12.1 The derivation of entropy as a property of System Z

132

We now suppose that System Z is taken from an initial Stable State 1 to a second Stable State 2 during which its temperature changes from T_1 to T_2 while, through reversible CHPPs, as required, it exchanges heat reversibly with the Datum Reservoir at constant temperature T_d. The Extended System Z^+ thus goes through a fully reversible process.

From the defining equation for thermodynamic temperature (equation 11.2), we have, for the CHPP,

$$\int_1^2 \left(\frac{dQ_T}{T}\right)_{\mathrm{REV}} = \int_1^2 \left(\frac{dQ_d}{T_d}\right)_{\mathrm{REV}} = \frac{[(Q_d)_{\mathrm{REV}}]_1^2}{T_d}. \tag{12.1}$$

Now, from the First Reversible-Work Theorem, not only is the gross work $[(W_g)_{\mathrm{REV}}]_1^2$ the same for all such fully reversible processes between the specified Stable States 1 and 2, but so also is $[(Q_d)_{\mathrm{REV}}]_1^2$, as we saw in Section 10.5. Thus, if we write

$$dS \equiv \left(\frac{dQ_T}{T}\right)_{\mathrm{REV}}, \tag{12.2}$$

then we have

$$(S_2 - S_1) = \frac{[(Q_d)_{\mathrm{REV}}]_1^2}{T_d} \tag{12.3}$$

and we see that the change in S, which is dependent only on Q_T and T, will be the same for *all* internally reversible processes by which System Z may be taken from Stable State 1 to Stable State 2. In 'Thermotopia', where all processes would be reversible, S would thus fulfil the requirement of a *property* of the system, as set out in paragraph (a) of Section 1.14, and we call this property 'the *entropy* of the system'. This means that the change in entropy of a system between two Stable States must be a function only of other thermodynamic properties of the system, a conclusion which is of importance when we come to discuss in the next Section what meaning we can attach to the change in entropy of a system when the process bringing about the change of state is irreversible.

That S also meets the requirement of a property that it should satisfy equation (1.2) in paragraph (b) of Section 1.14 can be seen from the Second Reversible-Work Theorem of Section 10.8. Equation (12.1) is still valid for the special case in which, from an initial Stable State 1, System Z is taken back to this state by a completed cyclic process, in which case we have

$$\oint \left(\frac{dQ_T}{T}\right)_{\mathrm{REV}} = \frac{(Q_d)_{\mathrm{REV\ CYCLE}}}{T_d}. \tag{12.4}$$

Now, from the Second Reversible-Work Theorem, $(Q_d)_{\mathrm{REV\ CYCLE}} = 0$, so that from equations (12.2) and (12.4) we have

$$\oint dS \equiv \oint \left(\frac{dQ_T}{T}\right)_{\text{REV}} = 0,$$ (12.5)

which is in accordance with equation (1.2) relating to any property P.

Our new thermodynamic property, entropy S, may thus be defined formally as follows:

Definition

When a system which is undergoing an internally reversible process takes in heat dQ_T at an instant when the thermodynamic temperature of the system is T, it suffers a change in *entropy* dS given by the relation

$$dS \equiv \left(\frac{dQ_T}{T}\right)_{\text{REV}}.$$ (12.6)

12.3 Change of entropy in an irreversible process between Stable States

Since entropy change is defined in terms of a reversible process, while all natural processes are in some measure irreversible, we need to discuss what, if anything, is meant by the entropy change that results from a change in state of a system that is brought about by an irreversible process. To answer this question, we first note that we established in Section 12.2 that the entropy change *between two specified Stable States* of a system is dependent only on those end states and must therefore be a function only of other thermodynamic properties of the system in those end states. Thus, insofar as the change in entropy of the system is concerned, it is immaterial whether the change of state is brought about by a reversible process or an irreversible process. In the case of a reversible process, in addition to the system being in a Stable State at the beginning and end of the process, it will pass through a continuous succession of Stable States. In an irreversible process, it will not pass through such a succession of Stable States, but the term 'entropy change' will still be meaningful *if the system starts from an identifiable Stable State and, on settling down to an equilibrium condition on conclusion of the irreversible process, is again in an identifiable Stable State.* Furthermore, since, as we have just deduced, the entropy change between two Stable States is a function only of other thermodynamic properties of the system in those two states, *the entropy change of the system between two identifiable Stable States will be the same whether the intervening process is reversible or irreversible.* Hence, for an irreversible process between Stable States 1 and 2, we evaluate the entropy change ΔS by applying equation (12.6) to *any alternative reversible process* between the two states, so that

$$[\Delta S_{\text{IRREV}}]_1^2 \equiv S_2 - S_1 = \int_1^2 \left(\frac{dQ_T}{T}\right)_{\text{REV}}.$$ (12.7)

12.4 Some simple applications of entropy

12.4.1 Temperature–entropy diagram

In Fig. 12.2(a), the path 1–2 shows the change in temperature of a system with change in entropy when the system is taken from Stable State 1 to Stable State 2 by an internally reversible process during which heat is supplied to the system. From the definition of entropy, equation (12.6), we have

$$(dQ_T)_{\mathrm{REV}} = T\,dS,$$

so that it is evident that the hatched area below path 1–2 in Fig. 12.2(a) is equal to the heat supplied in the complete process.

It should be particularly noted that if process 1–2 had been internally irreversible, the system would not have passed through a succession of Stable States, so that it would not have been possible to trace out the path on the temperature–entropy diagram. It is equally important to realize that if, as in Fig. 12.2(b), we then attempted to represent the process by a dotted line between the initial and final Stable States 1 and 2, if these were known, the area under this line would *not* represent the heat supplied to the system during the process.

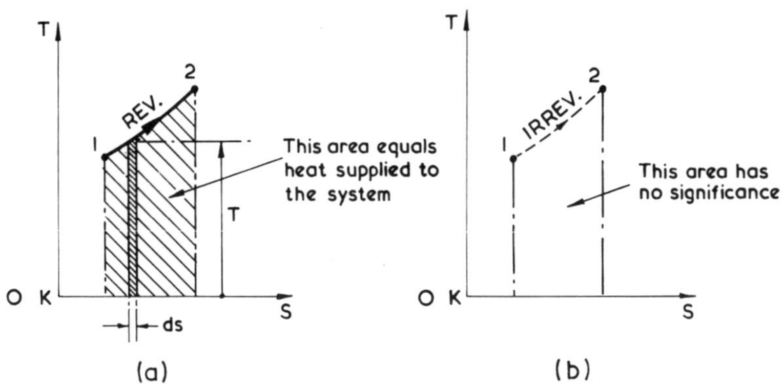

Fig. 12.2 Representation on a temperature–entropy diagram of a process between initial and final Stable States, when the process is (a) internally reversible, (b) irreversible

12.4.2 Reversible, adiabatic (isentropic) process

It follows from equation (12.6) that if a process is both internally reversible and adiabatic (i.e. work only, no heat), then the entropy of the system does not change during the process. We then describe the process as being *isentropic*.

Although isentropic processes, being reversible, would exist only in 'Thermotopia', they are of the greatest importance and practical utility. For example, the behaviour of a fluid as it passes through the nozzles of a turbine is predicted by first performing an idealized calculation in which the fluid is assumed to expand isentropically from the given initial state to the final exit pressure; the real behaviour is then assessed by applying to the calculated result a coefficient or efficiency factor determined from practical experience. However, before we can demonstrate this and the third valuable feature of entropy mentioned in Section 12.1 we shall need to introduce further theorems.

12.5 The $T\,dS$ equations for Simple Systems

We may use the definition of entropy to derive two very useful equations relating the value of the entropy change of a Simple System (Section 5.3) to the changes in the values of other thermodynamic properties. To this end, we consider the Simple System depicted in Fig. 12.3 as it undergoes an infinitesimal, internally reversible process between identifiable Stable States 1 and 2 while it performs displacement work $(dW_d)_{REV}$ and receives heat $(dQ_T)_{REV}$ when it is at temperature T. The system is at rest (or, more generally, we consider a frame of reference with respect to which the system is at rest) and the effects of external force fields (gravity, electricity, magnetism) are assumed to be negligible.

The Energy Conservation Equation (Section 7.1) for the system is then

$$dU = (dQ_T - dW_d)_{REV},$$

where

$$(dQ_T)_{REV} = T\,dS$$

and

$$(dW_d)_{REV} = p\,dV \quad \text{(see Section 3.4)};$$

whence, for this infinitesimal, internally reversible process, we have

$$(T\,dS)_{REV} = (dU + p\,dV)_{REV}. \tag{12.8}$$

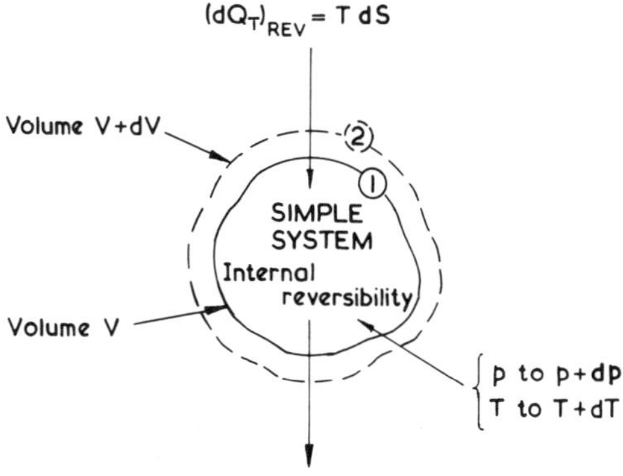

Fig. 12.3 Simple System undergoing internally reversible process

However, we noted in Section 12.3 that the entropy change of a system between two identifiable Stable States is the same whether the intervening process is reversible or irreversible. Hence the relation between dS, dU, and dV expressed by equation (12.8) is also applicable to an infinitesimal irreversible process *so long as the system starts from an identifiable Stable State and settles at the end of the process to a second identifiable Stable State*. We may thus drop the subscript REV and write

$$\boxed{T\,dS = dU + p\,dV} \tag{12.9}$$

as a thermodynamic relation which applies to changes between Stable States of a Simple System whether the process connecting them is reversible or irreversible. Nevertheless, in order to use equation (12.9) to evaluate the change in entropy for a finite change of state of a Simple System between identifiable Stable States, we are forced to consider a reversible process when integrating the equation, since only then is there an identifiable path along which the integration can be carried out.

A second very useful $T\,dS$ equation is obtained from equation (12.9) by making use of the defining equation for enthalpy (Section 7.8), namely

$$H \equiv U + pV,$$

to give

$$dH = dU + p\,dV + V\,dp; \tag{12.10}$$

whence, from equations (12.9) and (12.10), we have

$$\boxed{T\,dS = dH - V\,dp.} \tag{12.11}$$

We shall find these two $T\,dS$ equations invaluable when we come to consider the equilibrium thermodynamics of Simple Systems in Chapter 18, but here we shall simply seize the opportunity of using equation (12.11) to derive the result quoted in case (b) of Fig. 7.5 in relation to reversible shaft work in a steady-flow process.

12.6 Use of the second $T\,dS$ equation in the derivation of an expression for reversible shaft work in a steady-flow process

Figure 12.4 depicts diagrammatically an infinitesimal, internally reversible, steady-flow process in which there is assumed to be negligible difference between the kinetic energies of the fluid respectively entering and leaving the control volume and likewise negligible difference between the gravitational potential energies at these two stations. During passage through the control volume, each unit mass of fluid picks up heat of magnitude $(dQ_T)_{\text{REV}}$ and delivers shaft work of magnitude $(dW_x)_{\text{REV}}$.

Applying the Steady-Flow Energy Equation (equation 7.16) for unit mass of

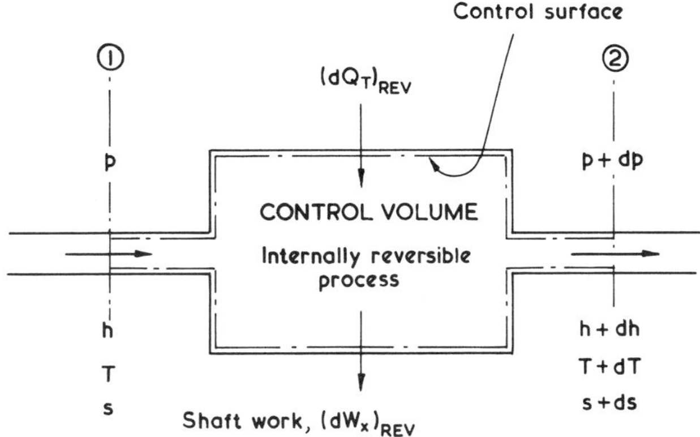

Fig. 12.4 Fluid undergoing infinitesimal, internally reversible, steady-flow process in passage through a control volume, with negligible changes in KE and PE. (*Note:* $(dQ_T)_{REV}$ and $(dW_x)_{REV}$ relate to *unit mass* of fluid as it passes through the control volume in steady flow. Thus, if $(\dot{Q}_T)_{REV}$ and $(\dot{W}_x)_{REV}$ are the *rates* of heat supply and work production respectively, and if $\dot{m}$ is the mass flow rate of fluid through a control volume, then, as in Section 7.11, $(Q_T)_{REV} = (\dot{Q}_T)_{REV}/\dot{m}$ and $(W_x)_{REV} = (\dot{W}_x)_{REV}/\dot{m}$.

fluid passing through the control volume in this infinitesimal process, we have

$$h + (dQ_T)_{REV} = (h + dh) + (dW_x)_{REV}$$

or

$$(dW_x)_{REV} = (dQ_T)_{REV} - dh. \tag{12.12}$$

We have so far studied only entropy changes of systems and have not yet considered how to apply the resulting expression to fluid flow through control volumes. We shall turn to this shortly in Section 12.8. In the meantime we can overcome this present shortfall in our knowledge by considering, in our imagination, a 'packet' of fluid of unit mass which maintains its individual separate identity as it passes through the control volume in steady flow. This is now our *system*, to which we can apply equation (12.6). While it is essentially at temperature T, this system picks up heat of magnitude $(dQ_T)_{REV}$ during passage through the control volume; thus, since the process is internally reversible, we shall have

$$(dQ_T)_{REV} = T\,ds. \tag{12.13}$$

A more generally acceptable method of arriving at this result will be presented in Section 12.8.

From equations (12.11), (12.12), and (12.13):

$$\boxed{(dW_x)_{REV} = -v\,dp} \tag{12.14}$$

138

or, for a finite change of state of the fluid,

$$(W_x)_{\text{REV}} = -\int_{p_1}^{p_2} v\, dp. \tag{12.15}$$

For positive shaft-work output the value of dp will clearly be negative; i.e. in the passage of the fluid through the device the pressure will fall. Particular attention is drawn to the difference between this expression and that for $(W_d)_{\text{REV}}$ in Section 3.4. The fact that equation (12.15) is valid whether or not the process is adiabatic indicates that it results from purely mechanical phenomena occurring within the control volume.

12.7 Thermal entropy flux

The concept of *thermal entropy flux* is one which we shall find of great utility. We introduce the concept by considering afresh the situation which gave rise to the definition of entropy, namely System Z of Fig. 12.1(a) as it undergoes an infinitesimal, internally reversible process during which it receives heat $(dQ_T)_{\text{REV}}$ when the system is at temperature T. We established in Section 12.2 that the increase in entropy of System Z is then given by

$$dS = \left(\frac{dQ_T}{T}\right)_{\text{REV}} \tag{12.2}$$

In seeking the source of this increase in entropy, we recognize that it certainly resulted from a flux of heat of magnitude $(dQ_T)_{\text{REV}}$ entering the system across a boundary which was at temperature T. It would therefore be appropriate, in this instance, to say that the entropy of the system had increased by dS as the result of the entry into the system of a *thermal entropy flux* of magnitude $(dQ_T/T)_{\text{REV}}$. That this will prove to be a useful concept remains to be demonstrated.

Since equation (12.2) is only valid for an internally reversible process, we have to ask ourselves whether the concept of (dQ_T/T) as a thermal entropy flux is of any utility when the process undergone by the system receiving the heat is internally irreversible. In Section 12.9 we shall, in fact, find that this is so. We therefore proceed to define the concept formally as follows:

Definition

When a system receives heat of magnitude δQ_T across a boundary which is at temperature T, the heat is said to bring into the system a *thermal entropy flux* of magnitude $(\delta Q_T/T)$.

Thus the following equation defines the thermal entropy flux dS_Q associated with dQ_T:

$$\boxed{\;\textit{Thermal entropy flux } dS_Q \equiv \frac{dQ_T}{T}.\;} \tag{12.16}$$

Moreover, this concept of thermal entropy flux into a system is useful when evaluating that part of the total increase in entropy of a system which is due to each of a number of heat transfers across different parts of the boundary surface of a system which are at different thermodynamic temperatures. This can be done so long as there are identifiable temperatures at these different locations; strictly speaking, this requires that, at each location, the part of the system concerned is in an identifiable Stable State during the heat transfer, since thermodynamic temperature is an equilibrium property. We shall then have a *surface* integral or summation, so that in these circumstances we can write

$$(\Delta S)_{\text{system}} = \sum_{\text{surface}} \frac{dQ_T}{T}. \tag{12.17}$$

12.8 Thermal entropy flux and convective entropy flux in flow processes

Having so far treated only the entropy change of a *system*, we now consider how to apply the knowledge so gained to fluid flow through a *control volume*. We may start by considering the infinitesimal, internally reversible, steady-flow process which was depicted in Fig. 12.4.

In Section 7.10, we arrived at the energy equation for a flow process by initially defining a system, to which we then applied the equation which we had already established for systems. We follow a similar procedure in the present context in order to arrive at the entropy equation for steady flow through the control volume of Fig. 12.4. Thus, as depicted in Fig. 12.5(a), we define a system comprising the contents of our control volume at a selected instant together with that unit mass of the fluid which is about to enter the control volume at location (1). We then consider that time interval during which this unit mass enters the specified control volume. Since we are considering a steady-flow process, at the end of that time interval a different unit mass of fluid will have left the control volume at location (2). The boundary of our same system is then as depicted in Fig. 12.5(b), no matter having crossed the boundary of the system between the instants depicted in (a) and (b). Thus in (b) the boundary of our system encloses the contents of the control volume together with the unit mass of fluid which has just left it.

We shall denote the entropy of the fluid within the *control volume* at instants (a) and (b) respectively by the symbols $(S_V)_a$ and $(S_V)_b$. Then, for our defined *system*, we shall have:

$$\text{Initial entropy of system} = (S_V)_a + s.$$

$$\text{Final entropy of system} = (S_V)_b + (s + ds).$$

Thus,

$$\text{Increase in entropy of system} = ds + [(S_V)_b - (S_V)_a].$$

Since the process is internally reversible, we shall thus have from equation (12.6):

$$\frac{(dQ_T)_{\text{REV}}}{T} = ds + [(S_V)_b - (S_V)_a]. \tag{12.18}$$

140

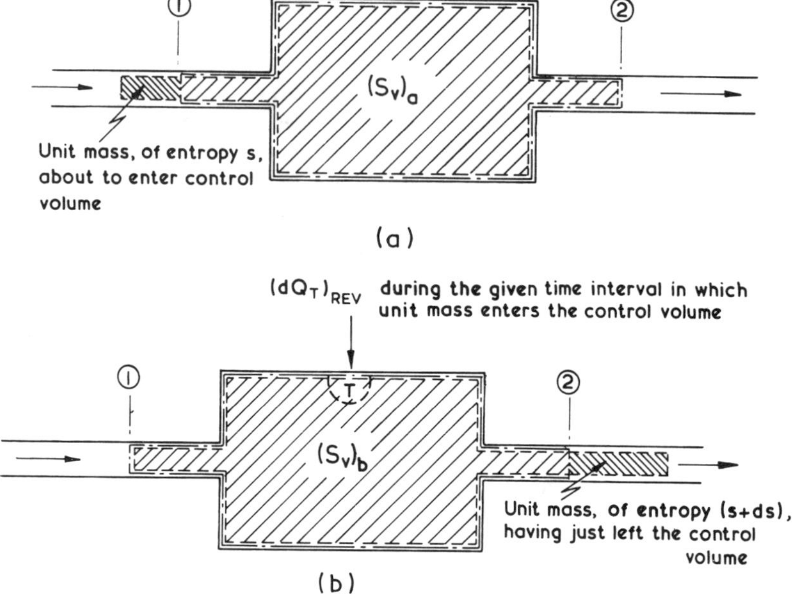

Fig. 12.5 Derivation of entropy equation for steady flow through a control volume. ––·–– is the control surface defining the control volume and ------ is the system boundary

Studying this equation in *relation to the control volume* of Fig. 12.4, we may now identify the three terms as follows:

$$\frac{(dQ_T)_{\text{REV}}}{T} =$$ *thermal entropy flux* into the *control volume* during the given time interval in which unit mass enters.

$$ds =$$ *net* efflux of *convective entropy* out of the *control volume* during the given time interval (namely the difference between the exit convective entropy flux $(s + ds)$ carried out by unit mass and the inlet convective entropy flux s brought in by unit mass).

$$[(S_V)_b - (S_V)_a] =$$ increase of entropy within the *control volume* during the given time interval.

We may thus describe equation (12.18) as the *Entropy Conservation Equation* for the *control volume* (**BUT** with a strong warning that, unlike energy, we only have a conservation equation for entropy when the process is internally reversible; we shall study that aspect next).

In steady flow the states of the fluid at all points within the control volume do not vary with the passage of time, so that $(S_V)_b = (S_V)_a$, and we then have

$$(dQ_T)_{\text{REV}} = T\,ds,$$

a result which we obtained as equation (12.13) in Section 12.6 by a more restricted line of argument.

12.9 Entropy conservation and entropy creation

With reference to the discussion in Section 12.7, we note that, when a process is internally reversible, the increase in entropy of a system is equal to the net thermal entropy flux into the system during the process. We may thus say that *in these circumstances* there is *entropy conservation* within the system. In Section 12.8, we have just recognized a corresponding condition of entropy conservation for flow through a control volume when the process is internally reversible. However, we shall show shortly in Section 12.11 that, when a process is internally irreversible, the entropy increase of the system is always *greater* than the net thermal entropy flux into the system during the process. We may therefore say that in these different circumstances there is *entropy creation* due to irreversibility, the magnitude of this entropy creation dS_c in a System Z whose entropy increases by dS_Z during an infinitesimal irreversible process between two closely adjacent Stable States being given by the *defining* expression

$$dS_c \equiv dS_Z - dS_Q = dS_Z - \frac{dQ_T}{T}. \tag{12.19}$$

For a finite irreversible process we may write this as

$$\boxed{\Delta S_c \equiv \Delta S_Z - \Delta S_Q = \Delta S_Z - \sum \frac{\delta Q_T}{T},} \tag{12.20}$$

where the summation is taken over all heat quantities at all points on the boundary of the system at which heat transfers occur, the temperature T being, in each case, *the particular temperature of that part of the boundary at which and when each heat transfer occurs.* We may thus formally define entropy creation in the following terms:

Definition

The *entropy creation* ΔS_c due to irreversibility within a system is that part of the entropy change of the system which cannot be accounted for by the thermal entropy fluxes associated with the heat transfers that take place across the boundary of the system.

Similarly, in the case of steady-flow processes, it is that part of the net efflux of *convective entropy* from the control volume that cannot be accounted for by the *thermal entropy* fluxes associated with heat transfers to or from the control volume.

We may summarize the foregoing discussion concisely thus:

(a) *Entropy Conservation Equation.* In an internally reversible process,

$$(\Delta S_c)_{REV} = 0. \tag{12.21}$$

142

(b) *Entropy Creation Expression.* In an internally irreversible process,

$$(\Delta S_c)_{IRREV} > 0. \tag{12.22}$$

We shall establish the truth of expression (12.22) in Section 12.11.

12.10 Entropy creation due to irreversibility — A particular case

Before establishing the general truth of expression (12.22), it is useful to gain further insight into the concept of entropy creation due to irreversibility by treating a particularly simple case.

We shall consider the steady flow of heat at a constant rate $\dot{Q}$ along a thermally resistive rod whose ends are respectively held at constant temperatures T_1 and T_2, while the rod is thermally insulated along its length and the state of the rod does not alter. This is illustrated in Fig. 12.6(a). We will first analyse this situation without using the concept of thermal entropy flux and then by using it.

For purposes of analysis, we may imagine the rod to draw heat at a constant rate $\dot{Q}$ from a Thermal Reservoir at constant temperature T_1 and to reject heat at the same constant rate to a second Thermal Reservoir at constant temperature T_2, as depicted in Fig. 12.6(b). We shall call the rod System X and the combined system of rod and reservoirs System Y. We know from Sections 6.8 and 9.6 that the process occurring within the rod is internally irreversible, but, by definition of the hypothetical device which we have called a Thermal Reservoir, the processes occurring within the latter are always reversible. Thus the entropy of the lower-temperature reservoir will increase at a rate $\dot{Q}/T_2$ and that of the higher-temperature reservoir will decrease at a rate $\dot{Q}/T_1$, while that of the rod will not alter since it is in a steady state. Thus, since T_2 is less than T_1, the entropy of *System Y* will increase at a net rate given by

$$\dot{S}_Y = \frac{\dot{Q}}{T_2} - \frac{\dot{Q}}{T_1}.$$

Now, this entropy increase of System Y must have resulted from the irreversibility within it, since Y is an *isolated* system having no heat transfers across its boundary. However, the only irreversibility within System Y lies within the rod, since we

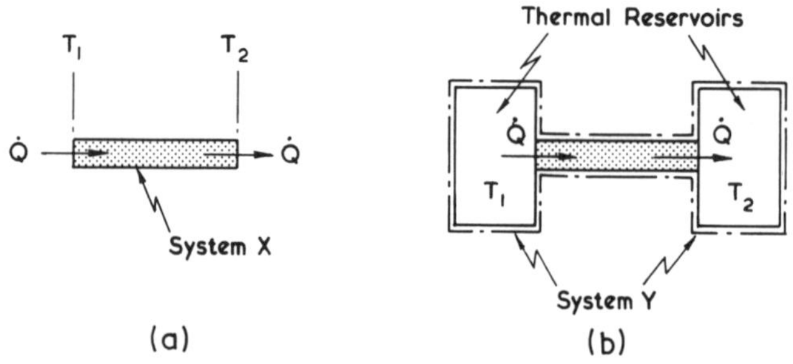

Fig. 12.6 Entropy creation during heat flow along a rod

know from the definition of a Thermal Reservoir that there are no irreversibilities within the two reservoirs. Thus we may say that the irreversibility within the rod has resulted in a rate of *entropy creation* $\dot{S}_c$ equal to $\dot{S}_Y$, namely

$$\dot{S}_c = \frac{\dot{Q}}{T_2} - \frac{\dot{Q}}{T_1}. \tag{12.23}$$

We will now derive the same result in a simpler fashion by applying, to the rod alone (System X), the concept of thermal entropy flux enunciated in Section 12.7. For System X we may re-write the defining equation for entropy creation, namely equation (12.20), in terms of time rates as follows:

$$\dot{S}_c = \dot{S}_X - \sum \frac{\dot{Q}_T}{T}, \tag{12.24}$$

where $\dot{S}_X = 0$ (the rod is in a steady state) and

$$\sum \frac{\dot{Q}_T}{T} = \frac{\dot{Q}}{T_1} - \frac{\dot{Q}}{T_2};$$

whence, from equation (12.24), we again obtain equation (12.23), namely

$$\dot{S}_c = \frac{\dot{Q}}{T_2} - \frac{\dot{Q}}{T_1}.$$

The gain in simplicity that is obtained by using the concept of thermal entropy flux is clearly evident in this example. Moreover, it is in the rod that the irreversibility resides and in which, therefore, the entropy creation due to irreversibility originates. Thus the second method, which needs to consider only the rod, is clearly more attractive than the first, which requires the addition of conceptual Thermal Reservoirs. We now turn to a presentation, in general terms, of this concept of entropy creation due to irreversibility.

12.11 General derivation of the concept of entropy creation due to irreversibility

That irreversibility, whatever its precise cause, always results in entropy creation may readily be demonstrated by making use of the First Reversible-Work Theorem of Section 10.4. With reference to Fig. 12.7(a), we suppose that System Z is taken by an internally irreversible process between two closely adjacent Stable States 1 and 2, during the course of which its temperature changes by an infinitesimal amount dT and its entropy increases by dS_Z, while it receives heat dQ_T across the boundary of the system at temperature T and delivers directly an amount of 'internal work' (as defined in Section 10.6) of magnitude dW_i. The entropy creation within System Z will then be given by

$$dS_c = dS_Z - \frac{dQ_T}{T}.$$

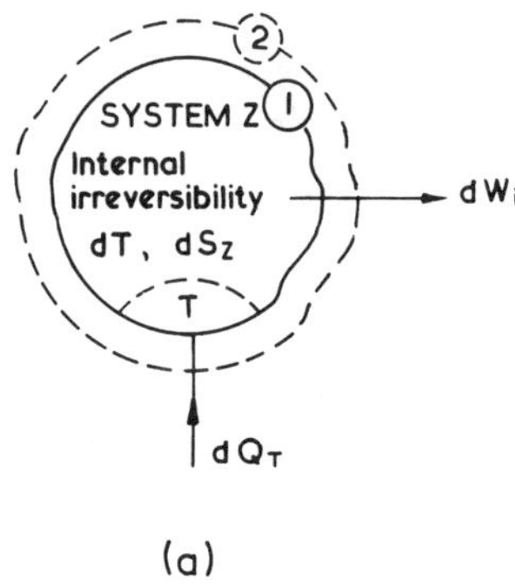

(a)

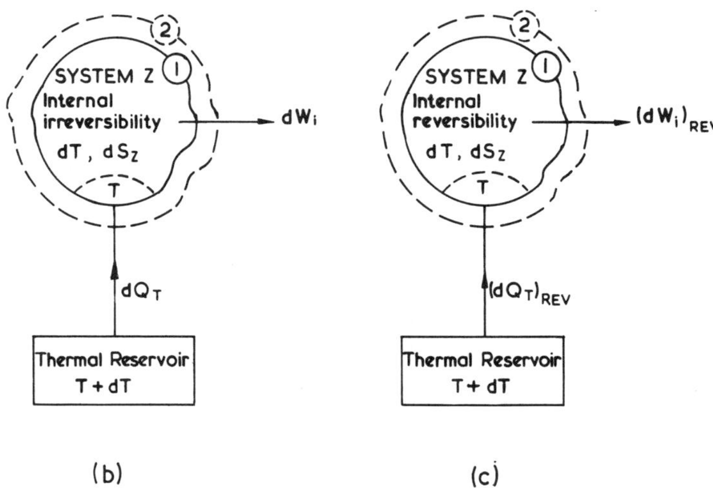

(b)   (c)

Fig. 12.7 General derivation of the concept of entropy creation due to irreversibility

Our task is now to show that this is necessarily always positive when the process undergone by System Z is irreversible.

As in the last Section, for purposes of analysis we may again imagine dQ_T to be supplied from a Thermal Reservoir, this time of temperature $T + dT$, as depicted in Fig. 12.7(b). There being no need for any auxiliary cyclic devices to maintain external reversibility in this transfer of heat to the system, the gross work output dW_g is in this case equal to the internal work dW_i.

Had System Z been taken by an alternative, hypothetical internally reversible process *between the same two Stable States 1 and 2*, its entropy change would have remained unaltered at dS_Z, but the heat received would then have been $(dQ_T)_{REV}$ and the internal work produced directly by the system would have been $(dW_i)_{REV}$. We can again imagine that, as depicted in Fig. 12.7(c), $(dQ_T)_{REV}$ is taken in from a similar Thermal Reservoir to that of Fig. 12.7(b). There being again no need for any auxiliary cyclic devices to maintain external reversibility in this transfer of heat to the system, the gross work output $(dW_g)_{REV}$ is equal to the internal work $(dW_i)_{REV}$. For system Z in this process, we also have the relation

$$dS_Z = \frac{(dQ_T)_{REV}}{T}.$$ (12.25)

Since System Z is taken between the same Stable States in both the irreversible process of Fig. 12.7(b) and the fully reversible process of Fig. 12.7(c), the Energy Conservation Equation for the system gives

$$(dQ_T)_{REV} - dQ_T = [(dW_i)_{REV} - dW_i] = [(dW_g)_{REV} - dW_g].$$ (12.26)

However, we know from the First Reversible-Work Theorem that $(dW_g)_{REV} > dW_g$, so that

$$(dQ_T)_{REV} - dQ_T > 0.$$ (12.27)

Now for System Z in the internally irreversible process of Fig. 12.7(a), the entropy creation due to irreversibility is given by

$$(dS_c)_{IRREV} = dS_Z - \frac{dQ_T}{T}$$

$$= \frac{(dQ_T)_{REV}}{T} - \frac{dQ_T}{T},$$

from equation (12.25). Hence, from this result and expression (12.27) we have

$$(dS_c)_{IRREV} > 0,$$

or, for a finite process between identifiable stable end states,

$$(\Delta S_c)_{IRREV} > 0.$$ (12.22)

As a corollary of the First Reversible-Work Theorem we have thus established the general truth of expression (12.22), which tells us that, in an internally irreversible process, the entropy creation is always positive. In Chapter 15 we shall show how this entropy creation can be used in a simple way to assess the loss of work output (or extra work input) due to irreversibility in work-producing and work-absorbing plant.

12.12 Special case of a thermally isolated system – The so-called 'Principle of the Increase in Entropy'

For a thermally isolated system there are no heat fluxes across its boundary, so that the second term (the heat-flux term) of expression (12.20) is zero. Furthermore, if such an isolated system Z undergoes an internally irreversible process, ΔS_c is positive and therefore ΔS_Z is also positive. Thus when, from an initial Stable State, a system passes through an irreversible adiabatic process to a final Stable

146

State, the entropy of the system in the latter state must be greater than that in the initial state. The statement of this fact has been described as the so-called *Principle of the Increase in Entropy*, but it will be noted that we have established its truth in the course of the logical development of our subject, so that we do not need to state it as a 'Principle' (or 'Law').

The foregoing result may be stated succinctly as follows:

$$(\Delta S_c)_{isol} \geqslant 0.$$

$(\Delta S_c)_{isol}$ is zero when the process is internally reversible and is always greater than zero (namely positive) when the process is internally irreversible. The physical significance of this interesting phenomenon is discussed briefly in Chapter 15.

Since all natural processes are in some measure irreversible, it follows from the foregoing discussion that *in all real-life* **adiabatic** *processes the entropy of the system must always increase and never decrease*. Thus, if it is possible to execute an adiabatic process which will carry the system from Stable State 1 to Stable State 2, we know that the entropy of the system is greater in state 2 than in state 1; moreover, it is also evident that it will not be possible to find any adiabatic process which would carry the system back to state 1 from state 2. Reference was made to this fact in Section 4.2.

One will not infrequently find the above situation described in terms which state that, as a system settles down from some initial non-equilibrium state to a final Stable State in an irreversible, adiabatic process, the entropy of the system moves towards *a maximum*. In terms of the entropy of classical equilibrium thermodynamics, as defined by equation (12.6), this statement is of doubtful validity, for we do not have a continuous function from which the entropy could be evaluated at the intervening non-equilibrium states. The entropy of classical equilibrium thermodynamics is an *equilibrium* property whose change in value can only be calculated between Stable States, so that we are in error if we make pretence that we can determine its value in any non-equilibrium state of a system.

Although, in classical equilibrium thermodynamics, there is no sense in which entropy can be regarded as moving towards a mathematical maximum in the above situation, statistical thermodynamics does reveal that the entropy of a system in a Stable State is associated with a true maximum of a mathematical function which defines the 'statistical entropy'. Discussion of the subject is frequently confused by the fact that this entropy is usually given the same symbol as the equilibrium entropy of classical thermodynamics. Using the symbol S_s for the statistical entropy, this may be defined by the expression

$$S_S = -k \sum_i p_i \ln p_i,$$

where k is Boltzmann's constant and p_i is the probability of occurrence of the ith microstate in a given macrostate. The equilibrium entropy S is then the maximum value of S_s obtained by differentiating this function with respect to the p_i, subject to a given set of mathematical constraints. Expressed rather freely, this operation may be interpreted as indicating that the movement of S_s to a maximum corresponds to a movement of the macroscopic state of the system towards that particular one which has a maximum number of possibilities of realization at the microscopic level. This is the Stable State of classical equilibrium thermodynamics and the equilibrium entropy S is associated only with such a macrostate of the system. It will be evident from this extremely superficial discussion that the equilibrium entropy S cannot be described as itself tending towards a mathematical maximum value in this situation. What we can say is that, if an isolated system is in some Stable State in which it is subject to specified fixed Constraints and its entropy is S_1, and an internal Constraint is then relaxed without leaving any effect on the environment, then the new Stable State to which the isolated system will finally settle will have an entropy S_2 such that $S_2 > S_1$.

12.13 Some practical applications involving the adiabatic expansion and compression of a fluid

In order to illustrate the practical aspects of the rather abstract concept of entropy creation due to irreversibility, we will consider the following three engineering examples of the steady, adiabatic flow of a compressible fluid: (a) through a converging nozzle, (b) through a turbine, and (c) through a compressor. Since the processes are adiabatic there is no heat transfer between the fluid and its environment. Consequently, we know from Section 12.11 that, in all three cases, the entropy of the fluid must increase as it passes through the devices. This is because there will be entropy creation ΔS_c due to irreversibility, since all real physical processes are irreversible in some degree. Thus, in Fig. 12.8, $s_2 > s_1$ in all three cases. On the other hand, we know that, in the ideal situation that would exist in 'Thermotopia', the processes would have been both adiabatic and reversible and the entropy of the fluid would therefore have remained constant; all three processes would then be described as *isentropic*, with $s_{2s} = s_1$. With the aid of the diagrams of Fig. 12.8, we now briefly study these effects and decide how to compare the actual performance with the ideal. For an understanding of the enthalpy–entropy diagrams of Fig. 12.8, the reader should first study Section E.2 in Appendix E at the end of this chapter.

12.13.1 Nozzle velocity coefficient ϕ

With reference to Fig. 12.8(a), we will assume that the velocity V_1 is sufficiently low for the inlet kinetic energy $V_1^2/2$ to be negligible compared with the exit kinetic energy $V_2^2/2$. Since the process is adiabatic and there is no work output, the Energy Conservation Equation for control volume Y (the Steady-Flow Energy Equation of Section 7.11) gives

$$\frac{V_2^2}{2} = (h_1 - h_2) \equiv \Delta h, \tag{12.28}$$

148

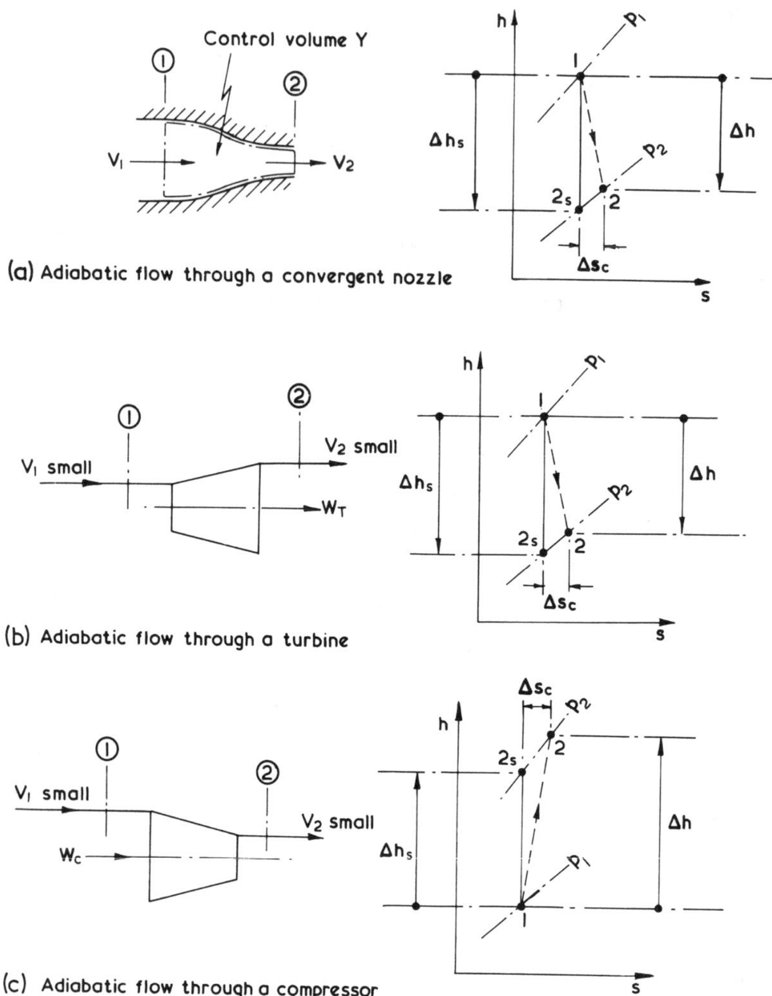

Fig. 12.8 Performance calculations for adiabatic steady-flow processes

where Δh is the *drop* in specific enthalpy of the fluid between inlet and outlet. Unhappily, we know neither the value of Δs_c nor that of Δh, so that we do not know the value of h_2 and are consequently unable to calculate V_2 from equation (12.28). We therefore set about the task of calculating V_2 by first considering an ideal, frictionless nozzle taking fluid at the same state 1 and expanding the fluid adiabatically and reversibly, at constant entropy, to state 2_s at the same exit pressure p_2. We may then determine the ideal enthalpy drop Δh_s from the known properties of the fluid and thence calculate the theoretical exit velocity V_{2_s} from

$$\frac{V_{2_s}^2}{2} = h_1 - h_{2_s} \equiv \Delta h_s. \tag{12.29}$$

Finally, we calculate V_2 from V_{2_s} by making use of an experimentally determined *velocity coefficient* ϕ, defined by

$$\phi \equiv \frac{V_2}{V_{2_s}}. \qquad (12.30)$$

It should be noted that, in the above calculation, it has been assumed that the velocity is uniform across a cross-section of the nozzle passage. In practice, owing to viscous effects in the fluid, the velocity will be lower near the wall than it is at the centreline; it is this fact that causes V_2 to be less than V_{2_s}.

12.13.2 Turbine isentropic efficiency η_T

With reference to Fig. 12.8(b), the velocities in the upstream and downstream ducts will be relatively low, so that we may assume that the difference between the inlet kinetic energy of the fluid to the turbine and the outlet kinetic energy will be negligible. For adiabatic flow through the turbine, the Steady-Flow Energy Equation will then give

$$W_T = (h_1 - h_2) \equiv \Delta h, \qquad (12.31)$$

where W_T is the shaft work produced by the turbine per unit mass flow of fluid flowing through it and Δh is the *drop* in specific enthalpy of the fluid between inlet and outlet.

Again, we know neither the value of Δs_c nor that of Δh. As in the case of the nozzle, we therefore set about the task of calculating Δh by first considering adiabatic, frictionless (isentropic) flow through an ideal turbine supplied with fluid at the same state 1 and exhausting at the same pressure p_2. Again, from the known properties of the fluid we can determine the ideal isentropic enthalpy drop Δh_s. We then relate Δh to Δh_s through an experimentally determined *isentropic efficiency* η_T for the turbine. This is a *component* efficiency, not to be confused with the thermal efficiency η_{CY} of a complete cyclic heat power plant of which the turbine may form a part. It is defined by the expression

$$\eta_T \equiv \frac{\Delta h}{\Delta h_s}. \qquad (12.32)$$

The actual turbine work output is then given by

$$W_T = \Delta h = \eta_T \, \Delta h_s. \qquad (12.33)$$

12.13.3 Compressor isentropic efficiency η_C

With reference to Fig. 12.8(c), the treatment of the compressor is very similar to that for the turbine, but one essential difference must be carefully noted. Whilst there is again an entropy increase Δs_c of the fluid due to irreversibility in passing through the compressor, we note that Δh, the *rise* in specific enthalpy of the fluid in the real compressor, is now *greater* than Δh_s. This means that the work input to the real compressor will be greater than that to an ideal compressor supplied with fluid at the same initial state 1 and delivering the fluid at the same exit pressure p_2; this conclusion is in accordance with the expectation that we have from our studies of the effects of irreversibility. We must therefore define the *isentropic efficiency* η_C for a compressor by the expression

$$\eta_C \equiv \frac{\Delta h_s}{\Delta h}. \tag{12.34}$$

The difference between equations (12.32) and (12.34) should be carefully noted.

After determining the ideal isentropic enthalpy *rise* Δh_s, the actual compressor work *input* is then calculated from

$$W_C = \Delta h = \frac{\Delta h_s}{\eta_C}, \tag{12.35}$$

using an experimentally determined value of η_C. Again, careful note should be taken of the difference between equations (12.33) and (12.35).

From the three foregoing examples, we see that the essentially thermodynamic calculations are made for 'Thermotopian' situations comparable to the real situation, but with all processes reversible. The actual performance is then related to this ideal performance through a factor determined from prior experimental knowledge. This is the pattern of all our calculations in the application of equilibrium thermodynamics to practical engineering problems.

12.14 Summary

In this chapter we have introduced the reader to one of the most difficult concepts in classical equilibrium thermodynamics, that of *entropy* as a thermodynamic property of a system. Recognizing that the First Reversible-Work Theorem of Section 10.4 provided a clue to the discovery of entropy as a property, we used this theorem to show that, when heat $(dQ_T)_{\text{REV}}$ is supplied to a system during an infinitesimal, *internally reversible* process while the system is at a thermodynamic temperature T, the quantity $(dQ_T/T)_{\text{REV}}$ is the same for all such internally reversible processes between the given initial and final Stable States. This quantity therefore represents the change in value of a property of the system, namely its change in entropy dS. We then debated whether we could attach any meaning to the entropy change of a system when the change in state of the system is brought about by an *irreversible* process. We concluded that, provided that the initial and final states of the system were identifiable Stable States, it was legitimate to evaluate the entropy change brought about by the irreversible process by calculating the entropy change for any alternative reversible process between the known initial and final Stable States.

We next demonstrated two particularly valuable features of this new thermo-

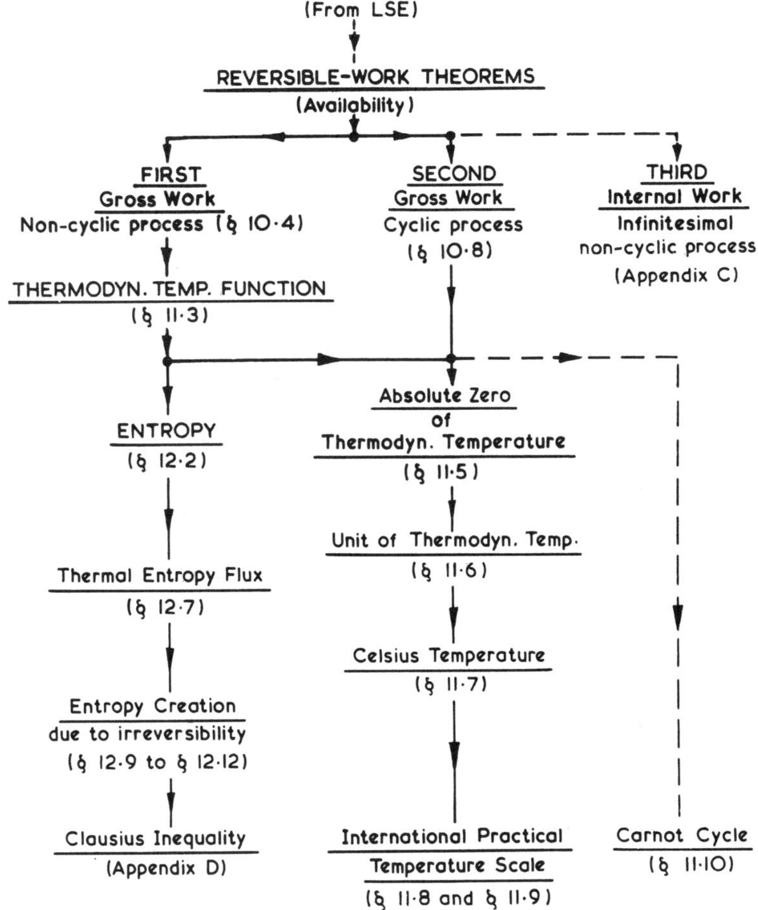

Fig. 12.9 Further development of the Thermodynamic Family Tree in Chapters 11 and 12 (continuation from Fig. 10.8)

dynamic property. First, it was pointed out that, if we plot the state path of an internally reversible process on a diagram with entropy of the system as abscissa and its thermodynamic temperature as ordinate, the area below the state path is equal to the heat supplied to (or extracted from) the system during the process. Second, that it followed from the definition of entropy that, in a process which was both adiabatic and internally reversible, the entropy of the system would remain constant during the process. Such a process was therefore described as being *isentropic*. The definition of entropy was then shown to lead to two particularly useful $T\,dS$ equations for a Simple System; one of these was used in the derivation of an expression for reversible shaft work in a steady-flow process.

After introducing the concept of a *thermal entropy flux* dS_Q, brought into or carried away from a system whenever heat dQ_T crosses the boundary of the system at a point where the temperature is T, we then introduced the further concept of

152

convective entropy flux in relation to processes involving flow through control volumes. These considerations led to the concept of an *Entropy Conservation Equation*, applicable only when a process was internally reversible, and to the very important concept of *entropy creation* due to irreversibility. We considered the latter in some detail, beginning with a particular case and leading on to the so-called 'Principle of the Increase in Entropy'.

The chapter ended with consideration of some practical applications involving the adiabatic expansion and compression of a fluid.

The material covered in this and the previous chapter is depicted in the further development of the Thermodynamic Family Tree shown in Fig. 12.9, which picks up from the point reached in Fig. 10.8. We shall not carry this Family Tree any further, since henceforth we shall be dealing principally with the development of the basic concepts treated in these last twelve chapters.

Appendix D at the end of this chapter presents the proof of a proposition called the *Clausius Inequality*, which features prominently in the method by which some texts introduce the concept of entropy. Appendix E presents further data relating to the thermodynamic properties of pure substances and perfect gases, supplementing the data given in Appendix A of Chapter 7.

APPENDIX D

The Clausius Inequality

Many texts deal first with a proposition called the *Clausius Inequality* before deriving the concept of entropy, whereas we have derived the latter without the aid of this proposition. Consequently, we now discuss the *Clausius Inequality* only in order that the reader can be made aware of its existence. Its derivation here follows simply from our concept of entropy creation due to irreversibility. As derived in these other texts, it is essentially a theorem in thermodynamic availability, though it has not hitherto been recognized as such.

We may rewrite equation (12.20) as

$$\sum \frac{dQ_T}{T} = \Delta S_Z - \Delta S_c,$$

and, in the particular case of a completed cyclic process, as

$$\oint \frac{dQ_T}{T} = (\Delta S_Z)_{\text{cycle}} - (\Delta S_c)_{\text{cycle}}.$$

If the cycle is reversible we have

$$(\Delta S_c)_{\text{cycle}} = 0.$$

Also

$$(\Delta S_Z)_{\text{cycle}} = 0.$$

Therefore,

$$\oint \left(\frac{dQ_T}{T}\right)_{\text{REV}} = 0. \tag{D.1}$$

If, on the other hand, the cycle is irreversible, we have

$$(\Delta S_c)_{\text{cycle}} > 0.$$

Also

$$(\Delta S_Z)_{\text{cycle}} = 0.$$

Therefore,

$$\oint \left(\frac{dQ_T}{T}\right)_{\text{IRREV}} < 0. \tag{D.2}$$

154

Hence, from equations (D.1) and (D.2):

$$\boxed{\sum \frac{dQ_T}{T} \leqslant 0.}$$

(D.3)

This is known as the *Clausius Inequality*. Equation (D.1) corresponds to equation (12.5) in Section 12.2.

Equation (12.20) may also be rewritten as

$$\Delta S_Z = \sum \frac{dQ_T}{T} + \Delta S_c.$$

Since ΔS_c is always positive, we shall have for a non-cyclic process:

$$\Delta S_Z \geqslant \sum \frac{dQ_T}{T}.$$

(D.4)

It must be noted that any use of these inequalities in relation to irreversible processes presupposes that, during any heat transfer to or from the system, it is possible to identify an equilibrium thermodynamic temperature T at that part of the system boundary across which the heat is being transferred.

APPENDIX E

Further Data on Pure Substances and Perfect Gases

The information presented in this Appendix supplements that appearing in Appendix A to Chapter 7. It adds to the diagrams given there two very useful diagrams involving the entropy of pure substances.* This Appendix also deals with the entropy of perfect gases.

E.1 Temperature–entropy diagram

We noted in Appendix A to Chapter 7 that we may represent the relation between any three thermodynamic properties of a pure substance by a three-dimensional surface. We may then draw two-dimensional diagrams by projecting features of this surface on to a chosen plane and use the projections of contour lines to show how the third property varies with variations in the other two.

The temperature–entropy diagram of Fig. E.1 represents the projection on the T–s plane of the T–s–p thermodynamic surface, since isobars (lines of constant pressure) are shown on the diagram. It is a particularly useful diagram for illustrating the processes occurring in power and refrigerating cycles using a condensable vapour. This is because, for ideal internally reversible cycles, the heat quantities supplied and rejected in the cycle may be represented by areas on the diagram. For this purpose those parts involving the solid phase are rarely required (an exception being carbon dioxide, whose triple-point pressure is above atmospheric, so that carbon dioxide 'snow', or 'dry ice', at atmospheric pressure sublimates directly to the gaseous phase without passing through the liquid phase).

Figure E.1 is based on the temperature–entropy diagram for water substance, though not being drawn to scale. It is not fully representative of pure substances in general. In particular, the slope of the saturated-gas line may be much more nearly vertical (and even positive) for fluids of large molar mass, such as the 'Freon' refrigerants; this may be seen from the p–h diagram for Refrigerant-12 (reference 8).

With reference to the isobars sketched in Fig. E.1, it may be noted that, if the substance in the gaseous phase could be treated as a perfect gas, then, from

---

* As noted in Appendix A, a pure substance is a somewhat idealized concept.

156

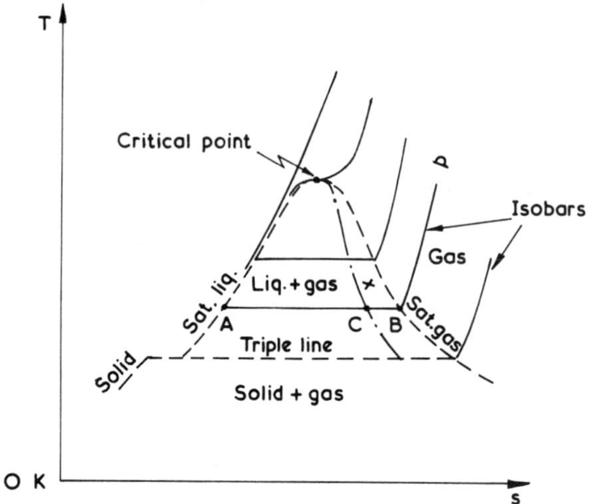

Fig. E 1 Temperature—entropy diagram for a pure substance

equation (12.11) of Section 12.5 and equation (A.9a) of Section A.10, we would have the relation

$$T\,ds = c_p\,dT - v\,dp.$$

The slope of an isobar would then be given by

$$\text{Slope} \equiv \left(\frac{\partial T}{\partial s}\right)_p = \frac{T}{c_p}. \tag{E.1}$$

Since c_p would be constant for a perfect gas, the slope of an isobar would thus increase with T. This effect can be seen in the isobars of Fig. E.1. The slope is zero in the two-phase regions because the saturation temperature is constant at a given saturation pressure.

E.2 Enthalpy—entropy diagram

The enthalpy—entropy diagram of Fig. E.2 is a particularly useful diagram for turbine design calculations, for which only a limited part of the diagram is needed.[14] Since isobars are plotted in Fig. E.2, which omits the solid phase, it represents the projection on the h–s plane of the h–s–p thermodynamic surface. It is based on the diagram for water substance.

For unit mass of a pure substance, it is readily deduced from the second $T\,ds$ equation, namely

$$T\,ds = dh - v\,dp, \tag{12.11a}$$

that, since

$$\left(\frac{\partial h}{\partial s}\right)_p = T, \tag{E.2}$$

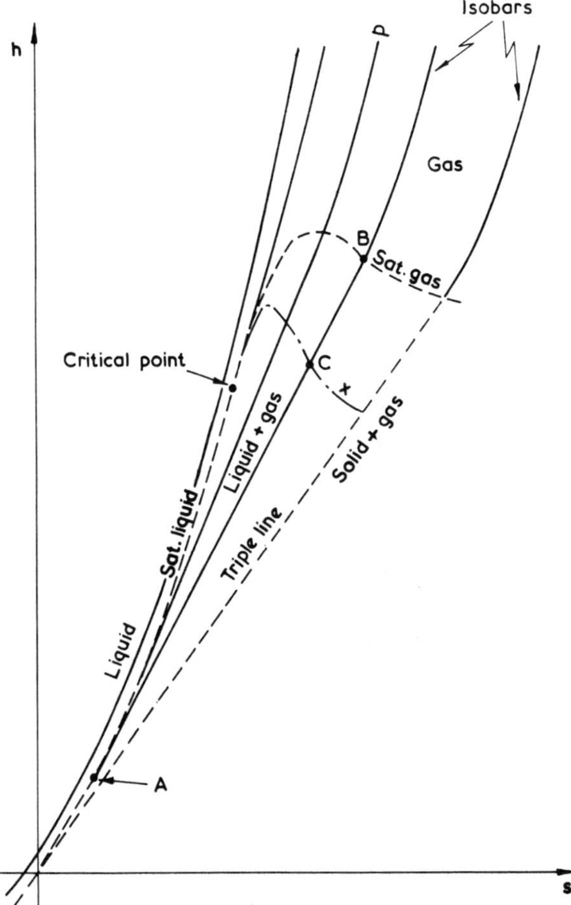

Fig. E.2 Enthalpy–entropy diagram for a pure substance (omitting the solid phase). [*Note:* The slope of an isobar at any state point on this diagram is equal to the thermodynamic temperature there; see equation (E.2).]

the slope of an isobar at any state point on this diagram is equal to the thermodynamic temperature at that point. Furthermore, isobars are straight lines in the two-phase liquid–gas region, the slope of a given isobar being equal to the saturation temperature corresponding to the given pressure.

E.3 Lines of constant dryness fraction in the liquid–gas two-phase region

The entropy S of a system is an extensive property and the entropies of constituent parts of a pure substance in a Stable State are directly additive if the

158

respective specific values are reckoned from the same datum state. Hence it is easily seen that, as in Figs. A.4 and A.6 in Appendix A, lines of constant dryness fraction x, such as the typical lines sketched in Figs. E.1 and E.2, are readily constructed by noting that

$$x = \frac{AC}{AB}.$$

(E.3)

E.4 Perfect gases

E.4.1 Change of entropy

In Section A.10, expressions were given for the changes in u and h of perfect gases; here we complete the picture by deriving alternative expressions for the change in entropy between Stable States.

We defined a perfect gas as one whose $p-v-T$ equation of state is

$$pv = RT$$

(A.7)

and whose specific heat capacities are constant. We then noted that, for any pure substance that has this equation of state, u and h are functions only of temperature (as we shall show mathematically in Section 18.12.1). We also have the following expressions, which are valid *for perfect gases* whether or not v and p are constant during a process:

$$du = c_v \, dT,$$

(A.8a)

$$dh = c_p \, dT.$$

(A.9a)

To obtain alternative expressions for the change in entropy of a perfect gas between two states, we may now use the two $T dS$ equations of Section 12.5, which, when expressed per unit mass, are written

$$T ds = du + p \, dv,$$

(12.9a)

and

$$T ds = dh - v \, dp.$$

(12.11a)

From equations (12.9a) and (A.8a) we have

$$ds = c_v \frac{dT}{T} + \frac{p}{T} dv,$$

whence, on substituting for p/T from equation (A.7), we obtain

$$ds = c_v \frac{dT}{T} + R \frac{dv}{v},$$

(E.4a)

or

$$s_2 - s_1 = c_v \ln \frac{T_2}{T_1} + R \ln \frac{v_2}{v_1}.$$

(E.4b)

Alternatively, from equations (12.11a) and (A.9a), we similarly obtain

$$ds = c_p \frac{dT}{T} - R \frac{dp}{p} \tag{E.5a}$$

or

$$s_2 - s_1 = c_p \ln \frac{T_2}{T_1} - R \ln \frac{p_2}{p_1}. \tag{E.5b}$$

A third alternative expression for the change in entropy of a perfect gas may be obtained from either equation (E.4a) or equation (E.5a), by noting from equation (A.7) that, since R is constant,

$$\frac{dT}{T} = \frac{dp}{p} + \frac{dv}{v}. \tag{E.6}$$

Also, we showed in Section A.10 that

$$R = c_p - c_v. \tag{A.12}$$

Hence, by combining equations (E.6) and (A.12) with either equation (E.4a) or equation (E.5a), we obtain

$$ds = c_v \frac{dp}{p} + c_p \frac{dv}{v} \tag{E.7a}$$

or

$$s_2 - s_1 = c_v \ln \frac{p_2}{p_1} + c_p \ln \frac{v_2}{v_1}. \tag{E.7b}$$

Equations (E.4a), (E.5a), and (E.7a), but not (E.4b), (E.5b), and (E.7b), are valid whether or not c_v and c_p are constant; they are thus equally valid for semi-perfect gases, as defined in Section A.9. The latter are discussed further in Chapter 17.

E.4.2 Reversible, adiabatic (isentropic) process

A reversible, adiabatic process is one of constant entropy, namely isentropic (Section 12.4.2), so that by putting $ds = 0$ in equations (E.7a), (E.4a), and (E.5a) respectively, it is easy to show that, during a reversible adiabatic expansion or compression of a perfect gas,

$$pv^\gamma = \text{constant}, \tag{E.8}$$

$$Tv^{\gamma-1} = \text{constant}, \tag{E.9}$$

$$\frac{T}{p^{(\gamma-1)/\gamma}} = \text{constant}, \tag{E.10}$$

160

where

$$\gamma \equiv \frac{c_p}{c_v}, \tag{A.13}$$

noting that

$$R = c_p - c_v. \tag{A.12}$$

Problems on Part I

Chapter 3. Work

3.1 Use the definition of *work* given in Section 3.2 to establish whether the net transfer of work from the specified system is positive, negative, or zero in the following examples:

(a) An electric locomotive is connected to a mains supply through frictionless overhead wires. The locomotive, with a truck attached, coasts downhill at constant speed without the brakes being applied and without the wheels slipping, while the motor is used as a generator to feed current into the mains. The track is perfectly rigid. Treat as the system: (1) the locomotive, (2) the truck, and (3) the locomotive and truck together.

(b) Repeat (a) when the locomotive uses a storage battery instead of being connected to the mains, so that when coasting downhill the battery is recharged.

Answer: See page 175.

3.2 State, giving reasons, whether the transfer of work from the specified system is positive, negative, or zero in the following examples:

(a) A mass (the system) falls in vacuo in a gravitational field.
(b) A man (the system) walks up a rigid staircase at constant speed.
(c) A man (the system) walks at constant speed up an upward-moving escalator: (1) for axes relative to the ground and (2) for axes relative to the escalator.
(d) A metal toboggan (the system) slides over a level sheet of ice, treated as being frictionless, and then on to a level patch of rough ground, where it coasts to a halt. (Treat the ground as rigid and thermally non-conducting and take axes relative to the ground.)
(e) A wooden toboggan (the system) slides over a level sheet of ice, treated as being frictionless, and then on to a fixed, level, rough sheet of metal, on which it coasts to a halt. (Treat the toboggan as rigid and thermally non-conducting and take axes relative to the ground.)

Answer: See page 175.

Chapter 4. The First 'Law' of Thermodynamics

4.1 Show, as a corollary of the Law of Stable Equilibrium, that it is impossible to execute a process involving no other effect than the lowering of a weight.

4.2 Prove Corollary 1 of the Law of Stable Equilibrium for adiabatic processes which involve work output from the system.

(For a problem demonstrating the significance of Corollary 1, see Problem E.2 of Appendix E to Chapter 12.)

Chapter 5. Energy

5.1 For the examples in Problems 3.1 and 3.2, indicate whether the respective changes in the energy E and the internal energy U are positive, negative, or zero for each of the specified systems.

In each case, describe the nature of the energy transfers during the process in question.

Answer: See page 175.

5.2 A rigid, insulated vessel having a volume of 1 m^3 contains 17 kg of steam. The contents of the vessel thus constitute a Constrained System subject to the specified fixed Constraint of confinement within a fixed bounding surface. By reference to Thermodynamic Tables[4] of the properties of steam, it may be verified that, in accordance with the State 'Principle', specification of the internal energy of the contents enables the Stable State of the system to be identified. Confirm this statement by considering, in turn, the following alternative values of the internal energy U of the contents of the vessel: (a) 44.24 MJ, (b) 46.36 MJ. In each case, specify the condition of the steam, listing its pressure and temperature.

Answer: (a) Dry saturated, 3.4 MN/m^2, 241 °C.
        (b) Superheated, 4.0 MN/m^2, 300 °C.

Chapter 6. Heat

6.1 Indicate whether Q, ΔE, and ΔU are positive, negative, or zero for the systems defined by italics as they undergo the following processes:

(a) *Air* in a cylinder is compressed by a piston and gets warm. Both the piston and cylinder are non-conducting.
(b) *Water* (1) falls over a waterfall and then (2) suddenly loses its kinetic energy at the bottom as it falls on a rigid, non-conducting rock.
(c) A *mixture* of hydrogen and oxygen in a non-conducting, rigid vessel suddenly burns spontaneously.
(d) The same as (c), except that the vessel is conducting.

(e) A *missile* is brought to rest by causing it to compress *air* behind a piston. The missile is rigid and all materials are non-conducting. Treat the two items separately and also as one system.

(f) A *lead–acid accumulator*, discharging adiabatically, supplies current to a *resistor*, whose temperature is kept constant. Treat the two items separately and also as one system.

Answer: See page 176.

Chapter 7. Work, Heat, and Energy – Energy Conservation Equations

7.1 A cork of negligible thermal conductivity is tied to the bottom of a large vacuum flask containing oil. The material of the flask is also of negligible thermal conductivity. The cork is released and eventually comes to rest on the surface of the oil. Write down, with explanation, the signs of the heat, work, and changes of energy, internal energy, and potential energy during this process for the systems respectively comprising (a) the contents of the flask, (b) the oil, and (c) the cork. Heat supplied *to* a system and work done *by* it are to be treated as being positive.

Repeat the exercise for a second process in which an inextensible string attached to the cork is threaded through an eye at the bottom of the flask and an external agent, pulling on the string, draws the cork slowly back to the bottom of the flask.

Answer: See page 176.

7.2 A wire is kept at ambient temperature while it is stretched by 1 m by a force which rises from zero to 1000 N, the extension of the wire being proportional to the force applied. Its length is then held constant while it receives 10 kJ of heat and its tension falls to zero. The wire is finally released and is returned at zero tension to its original state while rejecting 11 kJ of heat. Determine the magnitudes of W, Q, and ΔU for the wire during each of the three processes.

Answer: With all quantities expressed in kJ:

	W	Q	ΔU
Stretching	-0.5	0.5	1.0
Heating	0	10	10
Cooling	0	-11	-11

7.3 A substance is enclosed in a non-conducting cylinder by a conducting piston, the back of which is exposed to the pressure of the atmosphere ($0.102\ \mathrm{MN/m^2}$) but is thermally insulated from it. Friction between the cylinder and the piston ensures that the motion is slow as the substance expands from $1\ \mathrm{MN/m^2}$ and $1\ \mathrm{m^3}$ to $0.2\ \mathrm{MN/m^2}$ and $3\ \mathrm{m^3}$, and the process is represented by a straight line on the $p\!-\!V$

diagram. Neglecting the thermal capacity of the piston, calculate the work done by the substance, its change in internal energy, and the heat transfer to the substance.

Answer: 1.2 MJ, −0.204 MJ, 0.996 MJ.

7.4 A closed cylinder, thermally insulated from the environment except at one end, is divided into two parts by a light, frictionless, non-conducting piston. The cylinder has its axis vertical. That part of the cylinder above the piston is initially of volume 0.3 m^3 and contains air at a pressure of 7 kN/m^2 and a temperature of 39 °C. The lower part of the cylinder contains 0.05 kg of saturated water, whose volume may be neglected. Through the uninsulated lower end of the cylinder, heat is supplied slowly to the water until its pressure has risen to 28 kN/m^2.

Determine the final temperature of the air, the amount of water evaporated, and the quantity of heat supplied. It may be assumed that, during the compression process of the air, its temperature is related to the pressure by the expression $T/p^{(\gamma-1)/\gamma} = $ constant, where $T = (t + 273.15)$ kelvins and t is the Celsius temperature. (See Section E.4.2.)

Answer: 190.7 °C, 0.0338 kg, 82.3 kJ.

7.5 Natural gas is flowing from a large subterranean source at a depth of 3000 m below sea-level. Small evacuated storage tanks at sea-level are being filled slowly from the pipeline and the gas in each tank is found to be at 180 °C after filling. Estimate the temperature of the gas (a) entering the tanks and (b) in the subterranean source, assuming that all heat losses are negligible and that the gas is a perfect gas for which the molar mass is 16 kg/kmol and c_p/c_v is 1.3. Take the value of the gravitational acceleration to be 9.81 m/s^2 throughout the length of the pipe.

Answer: (a) 75.4 °C, (b) 88.5 °C.

7.6 An elastic balloon of negligible initial volume is filled from a compressed-air main through a throttle valve. The pressure and temperature of the air in the main are not affected by the withdrawal of air to fill the balloon, the temperature of the air in the main being 15 °C. The gauge pressure of the air in the balloon is directly proportional to the inflated volume, being 7 kN/m^2 when the balloon is inflated to a volume of 0.03 m^3. Calculate the temperature of the air in the balloon when it has been inflated slowly to this volume. The pressure of the atmosphere is 100 kN/m^2. Neglect heat transfer between the material of the ballon and the air inside.

Answer: 18.6 °C.

7.7 A boiler consists of an evaporator section and a superheater section. During a certain time interval, 5 kg of water at 15 MN/m^2 and 50 °C enter the boiler and a quantity of superheated steam at 15 MN/m^2 and 425 °C leaves it. The mass of saturated water in the evaporator section increases by 3 kg during this time interval. The pressure within the boiler remains constant at 15 MN/m^2, while both the total volume of the boiler and the volume occupied by the superheated steam in the superheater section also remain constant.

Calculate the decrease in the amount of saturated steam in the boiler and the amount of superheated steam which leaves it. Determine also the quantity of heat received by the boiler during the given time interval.

Answer: 0.481 kg, 2.481 kg; 10.09 MJ.

7.8 The power output of a steam turbine is 500 kW when steam flows through it at a steady rate of 5000 kg/h. Heat loss and turbine bearing loss are both negligible. Calculate the change in specific enthalpy of the steam as it flows through the turbine (a) when changes of kinetic energy and elevation between entrance and exit are negligible; (b) when the flow velocities at entrance and exit are respectively 50 m/s and 300 m/s, and the inlet pipe is 5 m above the exhaust.

Answer: (a) -360 kJ/kg, (b) -403.7 kJ/kg.

7.9 Steam at 1 MN/m^2 and $300\,^\circ$C approaches a horizontal nozzle through a passage of very large cross-section. Its pressure and temperature at exit from the nozzle are 0.5 MN/m^2 and $250\,^\circ$C. Neglecting heat loss from the nozzle, determine (a) the velocity at exit and (b) the area of the exit cross-section if the flow rate is 5 kg/s.

Answer: (a) 426 m/s, (b) 55.5 cm^2.

7.10 Steam flows at a rate of 10 kg/s through a horizontal pipe which has a constant cross-sectional area of 100 cm^2. The pressure and temperature of the steam are respectively 4 MN/m^2 and $500\,^\circ$C at inlet and 2 MN/m^2 and $450\,^\circ$C at exit. Determine the rate of heat loss from the steam and the horizontal force exerted by the pipe on its points of attachment.

Answer: 774 kW, 19.2 kN.

7.11 Calculate the amount of heat which enters or leaves 1 kg of steam, initially at 0.5 MN/m^2 and $250\,^\circ$C, when it undergoes the following processes:

(a) It is confined by a piston in a cylinder and is compressed to 1 MN/m^2 and $300\,^\circ$C as the piston does 200 kJ of work on the steam.
(b) It passes in steady flow through a plant and leaves at 1 MN/m^2 and $300\,^\circ$C, while, per kilogram of steam passing through the plant, a shaft puts in 200 kJ of work. Changes in kinetic energy and potential energy are negligible.
(c) It flows into an evacuated rigid container from a large source which is maintained at the initial condition of the steam. Then 200 kJ of shaft work is transferred to the steam, so that its final condition is 1 MN/m^2 and $300\,^\circ$C.

Answer: (a) -130 kJ. (b) -109 kJ. (c) -367 kJ.

7.12 Show that, if there is negligible difference between the kinetic energies of the fluid upstream and downstream of a throttle valve or orifice placed in an insulated pipe, there will be no difference between the specific enthalpies of the fluid upstream and downstream of the valve or orifice. How does the specific enthalpy of the fluid vary as it passes through the device?

Appendix A (Chapter 7)

A.1 Wet steam at $1.4\,\mathrm{MN/m^2}$ is reduced in pressure by passing it through an insulated throttle valve to a region where the pressure is $0.05\,\mathrm{MN/m^2}$. Assuming that there is negligible difference between the kinetic energies upstream and downstream of the valve, determine the initial dryness fraction of the steam when its temperature downstream of the valve is found to be $125\,^{\circ}\mathrm{C}$.

Answer: 0.971.

A.2 If the fluid passing through the throttle valve in Problem A.1 were air instead of steam, and if air can be treated as a perfect gas, what would be the temperature upstream of the valve?

Answer: $125\,^{\circ}\mathrm{C}$.

A.3 At a constant pressure of $2.033\,\mathrm{MN/m^2}$, 1 kg of saturated liquid ammonia is confined in a cylinder beneath a weighted piston. How much heat must be supplied before the volumes of liquid and vapour are equal? How much more heat would be needed to raise the temperature to $150\,^{\circ}\mathrm{C}$?

Answer: 29 kJ, 1316 kJ.

A.4 Define the term *critical point* in relation to a fluid. A hollow, horizontally mounted cylinder of constant volume has transparent end faces and is made of non-conducting material. After initial complete evacuation, water is admitted until liquid partly fills the vessel at a pressure of 1 atm. The cylinder is then sealed and the contents are slowly heated by an internal electrical heating element of negligible thermal capacity. Explain why the liquid level to which the vessel is initially filled affects the direction in which the meniscus will move as the fluid is heated.

What fraction of the total volume of the vessel must initially be occupied by liquid if, as energy is supplied, the disappearance of the meniscus is to occur at a position intermediate between the top and bottom of the cylinder? Calculate the quantity of energy that would then need to be supplied to cause the meniscus just to disappear, when the volume of the vessel is such that it contains a mass of 1 kg. What would then be the pressure within the vessel?

Answer: 0.329, 1616 kJ, $22.12\,\mathrm{MN/m^2}$.

A.5 It can be shown that, for a fluid substance,

$$\left(\frac{\partial u}{\partial v}\right)_T = T\left(\frac{\partial p}{\partial T}\right)_v - p.$$

Hence show that, for an ideal gas, u is a function only of T and that $h, c_v,$ and c_p are also functions only of T.

A.6 Calculate the value of the isochoric specific heat capacity c_v of a perfect gas for which the isobaric specific heat capacity c_p is 0.520 kJ/kg K and the molar mass is 40 kg/kmol. Calculate also the value of γ.

Answer: 0.312 kJ/kg K, 1.667.

Chapter 8. The Second 'Law' of Thermodynamics

8.1 Attempt the Examples of Section 8.3 without referring to the answers given in Appendix B.

8.2 Clausius stated the Second 'Law' in the form: *It is impossible for a system working in a cycle to have as its sole effect the transfer of heat from a system at a low temperature to a system at a higher temperature.* Show that, if it were possible to execute such a process, then it would be possible to construct a PMM 2.

Chapter 9. Irreversibility and Reversibility

9.1 By demonstrating that, with the aid of the effacing process (Section 9.3), it would be possible to construct both a Non-cyclic PMM 2 (Section 8.2) and a Cyclic PMM 2 (Section 8.6), show that the following processes are irreversible:

(a) A weight is attached to a light string which is wrapped round the spindle of a flywheel, the spindle being mounted in frictionless bearings. A non-conducting, friction brake-block applied to the rim of the flywheel allows the weight to descend at constant velocity.

(b) Two vessels are connected by a pipe in which there is an isolating valve, the whole being perfectly insulated. The vessels initially contain air at the same temperature but at different pressures. The valve is then opened and the pressures are allowed to equalize. (The air may be treated as a perfect gas.)

(c) Heat transfer across a finite temperature difference.

(d) The combustion of gaseous hydrogen and gaseous oxygen in an adiabatic, steady-flow combustion chamber. It may be assumed that the two gases enter at the same temperature and pressure and that there is no pressure drop across the combustion chamber.

(e) The diffusive mixing in slow steady flow, at constant temperature and pressure, of a stream of gaseous argon and a stream of gaseous nitrogen. The two gases may be treated as being perfect and inert.

In the proof of (e), it will be necessary to make use of the conceptual devices known as *semi-permeable membranes*, discussed in Part II, Chapter 19.

9.2 In two steady-flow compressors, R and A, gas passes at the same rate between the same pair of states. Compressor R is reversible* and compressor A is adiabatic. Show that A cannot require *less* shaft work to drive it than does R.

If A requires *more* shaft work to drive it than does R, show that A is irreversible. (In each case, show that, if the statement to be proved were not true, then a PMM 2 could be constructed.)

Chapter 10. Single-reservoir Processes and Reversible-Work Theorems

10.1 A system which can exchange heat with a single Thermal Reservoir is taken between specified end states in a work-absorbing process. Show that, if the process is irreversible, the gross work input will be *greater* than if it is fully reversible.

(Problems 13.1 and 13.3 to Chapter 13 in Part II provide numerical confirmation of the truth of the First Reversible-Work Theorem for fully reversible processes.)

Chapter 11. Two-Reservoir Processes and Thermodynamic Temperature

11.1 A hypothetical cyclic heat power plant operates on the Carnot cycle between reservoir temperatures of 250 °C and 25 °C. Calculate the thermal efficiency of the plant.

Answer: 43.0 per cent.

11.2 Calculate the coefficient of performance of a hypothetical cyclic *refrigerator* plant operating on the reversed Carnot cycle between reservoir temperatures of 0 °C and 20 °C.

Answer: 13.7.

11.3 Calculate the coefficient of performance of a hypothetical cyclic *heat-pump* plant operating on the reversed Carnot cycle between reservoir temperatures of 15 °C and 30 °C.

Answer: 20.2.

11.4 The work output from the power plant of Problem 11.1 is used to supply the work input to the heat-pump plant of Problem 11.3. Calculate the ratio of the heat supplied to the upper-temperature reservoir of the heat-pump plant to that extracted from the upper-temperature reservoir of the power plant.

Answer: 8.7.

---

* Note that, when we say that a *device* is reversible or irreversible, we mean that the *process* which it executes is respectively reversible or irreversible.

11.5 Complete, where relevant, the following table relating to two-reservoir cyclic plant:

Case	1	2	3	4	5	6
Source temperature, °C	327		300	60	327	140
Sink temperature, °C	27	70	100	60	827	
Heat from source, kJ/h	10 000	7 000			6 000	6 000
Heat to sink, kJ/h	5 500			1 440		5 000
Power output, kW			29.3	−0.4		0.3
Thermal efficiency, %			40			
Coefficient of performance		3.5			2.2	
Reversible, irreversible, or impossible?		Rev.				
CHPP, refrigerator, or heat pump?						

Answer: See page 176.

11.6 Show that the coefficient of performance of an irreversible heat-pump plant operating between two Thermal Reservoirs of specified temperature will necessarily be less than that of a reversible heat-pump plant operating between the same two reservoirs.

11.7 A heat-pump plant takes in 600 J of heat at 300 K while rejecting heat to a system whose temperature rises by 1 K for each joule of heat supplied to it. The system is initially also at 300 K. Show that there is a limit on the final temperature of the system and calculate its value. Is it an upper or a lower limit and under what conditions would it be attained?

Answer: 2216.7 K, lower (when plant is reversible).

Chapter 12. Entropy

12.1 A copper rod is of length 1 m and diameter 0.01 m. Its ends are held at 100 °C and 0 °C respectively. The rod is perfectly insulated along its length and the thermal conductivity of copper is 380 W/m K. Calculate the rate of heat transfer along the rod and the rate of entropy creation due to the irreversibility of this heat transfer.

Answer: 2.985 W, 0.00293 W/K.

12.2 Air is compressed isothermally, at the environmental temperature of 20 °C, through a pressure ratio of 2 in a steady-flow process in which there is negligible difference between the kinetic energies of the air at inlet and exit. The work input to the compressor is 73 kJ per kilogram of air compressed.

Calculate, per kilogram of air compressed: (a) the amount by which the convective entropy flux at exit exceeds that at inlet, (b) the thermal entropy flux entering the air in its passage through the compressor, and (c) the entropy creation Δs_c due to the irreversibility of the compression process.

Calculate the work input for reversible isothermal compression of 1 kg of air through the same pressure ratio. Confirm that the actual work input exceeds this quantity by an amount equal to $T_0 \, \Delta s_c$, where T_0 is the thermodynamic temperature of the environment. (A general proof of this result appears in Chapter 15.)

Answer: (a) -0.1989 kJ/kg K, (b) -0.2490 kJ/kg K, (c) 0.0501 kJ/kg K; 58.3 kJ/kg.

12.3 State clearly the conditions under which the two $T \, dS$ equations of Section 12.5 are applicable.

A light string passes without slipping over a freely running pulley and carries a ball of mass 2 kg at one end and a ball of mass 1 kg at the other. The 2-kg ball is immersed in a large body of viscous oil at a temperature of 15 °C, while the 1-kg ball is immersed in a large body of water, which is also at 15 °C. The balls are made of non-conducting material whose specific gravity is 3.0; the specific gravity of the oil is 0.9. Stray heat transfer from the oil and water to the environment is negligible.

The 2-kg ball is descending slowly in the oil at constant velocity, the 1-kg ball remaining below the surface of the water. Calculate the change in entropy of (a) the oil and water together and (b) the compound system of balls and string, as the balls move through a distance of 1 m. What do these changes in entropy signify?

Answer: (a) 0.025 J/K; (b) 0.

12.4 At low temperatures, over a limited temperature range, the change in internal energy of water at atmospheric pressure may be taken as being given by the expression

$$du = 4.19 \, dT,$$

where u is in kilojoules per kilogram and T is in kelvins. Treating water under these conditions as being incompressible, write down an expression for its specific entropy in terms of temperature.

Calculate the overall entropy change when 1 kg of water at 127 °C is mixed adiabatically with 1 kg of water at 27 °C. What does this entropy change signify?

Answer: 0.0863 kJ/kg K.

12.5 Wet steam at 1.4 MN/m² is reduced in pressure by passing it through an insulated throttle valve to a region where the pressure is 0.05 MN/m². The temperature downstream of the valve is found to be 125 °C. Assuming that kinetic energies upstream and downstream of the valve are negligible, determine the initial dryness fraction of the steam. Determine also, per kilogram of steam flowing through the valve, the entropy creation due to irreversibility.

Answer: 0.971, 1.478 kJ/kg K.

12.6 A fluid suffers a small pressure drop in flowing adiabatically along a horizontal passage, there being negligible change in the kinetic energy of the fluid. Show that the change in specific entropy of the fluid is related to the change in pressure by the expression

$$\frac{ds}{dp} = -\frac{v}{T}.$$

In a steam power plant, steam leaves the boiler at 2 MN/m² and 350 °C and exhausts from the turbine to the condenser at 7 kN/m². There is a 5 per cent. pressure drop, and negligible heat loss, between boiler and turbine, and the expansion in the turbine is isentropic. Using the above expression, estimate the increase in specific entropy of the steam in passing from the boiler to the turbine.

Noting that wet steam enters the condenser, calculate directly, per kg of steam, the reduction in turbine work output resulting from this pressure drop between boiler and turbine. (*Hint:* The reduction in work output is equal to the increase in the heat quantity rejected in the condenser.) Express this reduction as a percentage of the isentropic work output that would have been obtained had there been no such pressure drop.

Answer: 0.0222 kJ/kg K; 6.9 kJ/kg, 0.7 per cent.

12.7 In passing through a compressor in steady flow, air is compressed reversibly until its density is doubled, its temperature at entry being 15 °C. Calculate the shaft-work input per kilogram of air compressed when the process is (a) adiabatic, (b) isothermal. Kinetic energies are negligible.

Answer: (a) 93.0 kJ/kg, (b) 57.3 kJ/kg.

12.8 The water passing through a water turbine in steady flow is in state 1 at inlet and in state 2 at outlet. Above an arbitrary datum level, the heights at these two recording positions are respectively z_1 and z_2. There is negligible difference between the kinetic energies of the water at inlet and outlet. The flow through the turbine may be treated as being adiabatic and incompressible.

By application of the Steady-Flow Energy Equation, show that the shaft work delivered per unit mass of water flowing through the turbine is given by

$$W_x = \frac{p_1^* - p_2^*}{\rho} - (u_2 - u_1),$$

where the manometric pressure $p^* \equiv p + \rho g z$ and ρ is the density of the water.

The isentropic efficiency of the turbine is defined as the ratio of the actual work

172

output to the ideal work output from a reversible turbine subject to the same difference in manometric head $\Delta H^* \ [\equiv (p_1^* - p_2^*)/\rho g]$. Show that the isentropic efficiency η_T is related to the temperature rise ΔT of the water in passing through the turbine by the expression

$$\eta_T = \left(1 - \frac{c_v \, \Delta T}{g \, \Delta H^*}\right),$$

where c_v is the isochoric specific heat capacity of the water.

When the difference in manometric head across such a turbine is 30 m, the measured temperature rise of the water is 0.015 K. Calculate the isentropic efficiency of the turbine, given that $c_v = 4.19$ kJ/kg K.

Answer: 78.6 per cent.

12.9 Steam enters a turbine nozzle at 1 MN/m^2 and 300 °C with negligible kinetic energy and leaves at 0.5 MN/m^2 and 225 °C. Neglecting heat loss from the nozzle, calculate the velocity at exit and the nozzle velocity coefficient. Determine also, per kilogram of steam flowing through the nozzle, the entropy creation due to irreversibility.

Answer: 535 m/s, 0.932; 0.044 kJ/kg K.

12.10 The following figures relate to a performance test on a steam turbine:

    Steam pressure at turbine inlet = 1.0 MN/m^2
    Steam temperature at turbine inlet = 300 °C
    Condenser pressure = 6 kN/m^2
    Turbine output = 5000 kW
    Rate of steam supply = 6.95 kg/s

Neglecting mechanical losses and stray heat loss, determine the drop in specific enthalpy of the steam in passing through the turbine and the isentropic efficiency of the turbine. Calculate also, per kilogram of steam flowing through the turbine, the entropy creation due to irreversibility.

Answer: 719.4 kJ/kg, 83.9 per cent.; 0.447 kJ/kg K.

12.11 An axial-flow air compressor has an isentropic efficiency of 84 per cent. when compressing air adiabatically from a temperature of 17 °C over a pressure ratio of 4.13. Determine the exit temperature of the air. Calculate also, per kilogram of air compressed, the required shaft-work input and the entropy creation due to irreversibility. The air may be treated as a perfect gas.

Answer: 189.7 °C; 174.4 kJ/kg, 0.0647 kJ/kg K.

Appendix D (Chapter 12)

D.1 An irreversible cyclic heat power plant produces work when exchanging heat with two Thermal Reservoirs at temperatures of 500 °C and 50 °C. The working fluid is taken through a practical approximation to a Carnot cycle which may be represented by a trapezium on the temperature–entropy diagram. There are entropy increases during both the adiabatic compression and the adiabatic expansion, each equal to 10 per cent. of the range of entropy variation during the cycle. The fluid may be assumed to change its state reversibly during the processes of heat exchange with the reservoirs, and each heat exchange takes place with a constant temperature difference of 50 K between the reservoir and the working fluid.

Calculate the thermal efficiency of the plant and also that of a Carnot cycle operating between the same reservoirs. Confirm that the cyclic process occurring in the actual plant satisfies the Clausius Inequality.

Answer: 35.5 per cent., 58.2 per cent.

D.2   A reversible cyclic heat power plant exchanges heat with three Thermal Reservoirs at temperatures of 400 K, 300 K and 200 K respectively. During an integral number of complete cycles the plant absorbs 1200 kJ from the reservoir at 400 K and performs 200 kJ of mechanical work. Calculate the quantities of heat exchanged with the other two reservoirs.

Answer: -1200 kJ, $+200$ kJ.

D.3   One kilogram of air undergoes a change of state from 5 atm and 900 K to 1 atm and 600 K. The only other final effects are an amount of heat Q_0 taken from the environment at 300 K and the production of a gross work output of magnitude W_g. Use expression (D.4) to determine the upper limits on Q_0 and W_g. Under what hypothetical circumstances could this upper limit on W_g be obtained?

Calculate the value of W_g if, in the actual process, Q_0 is equal to -10 kJ (that is, 10 kJ are delivered *to* the environment). Thence determine the loss of gross work output due to the irreversibility of the actual process. Confirm that this loss is equal to $T_0 \Delta S_c$, where ΔS_c is the entropy creation due to irreversibility, defined by equation (12.20).

Answer: 15.7 kJ, 231.7 kJ; 206.0 kJ, 25.7 kJ.

Under hypothetical circumstances of *full reversibility* (Section 10.6).

(For proof of the relation between loss of gross work output and entropy creation due to irreversibility, see Section 15.2 of Chapter 15 in Part II.)

Appendix E (Chapter 12)

E.1   Plot an enthalpy–entropy diagram for water and steam on which the following features are shown:

(a)   The saturation line for water and steam.
(b)   The critical point and the triple-point line.
(c)   Isobars at pressures of 0.01, 0.1, 1.0, and 50 MN/m². For any one of these isobars, confirm that its slope in the wet region is equal to the corresponding absolute saturation temperature.
(d)   Isotherms at temperatures of 100, 250, and 500 °C.
(e)   Lines of constant dryness fractions of 0.5 and 0.75.

174

(f) The phases present in each region.

(g) The processes in Problems 12.5, 12.6, 12.9, and 12.10.

Ranges: h from 0 to 4000 kJ/kg.

   s from 0 to 10 kJ/kg K.

Scales: 1 cm $\equiv$ 200 kJ/kg and 0.5 kJ/kg K.

E.2 Unit mass of air is confined within a cylinder by a piston to which an external force can be applied along the piston rod. Within the cylinder there is an externally driven paddle wheel by means of which work may be put into the air by a stirring process. The air is initially of volume v_1 in Stable State 1 at pressure p_1 and is then taken by an *adiabatic* process to a new Stable State 2 in which the volume is $2v_1$. The following alternative processes for bringing about this change of state are to be studied:

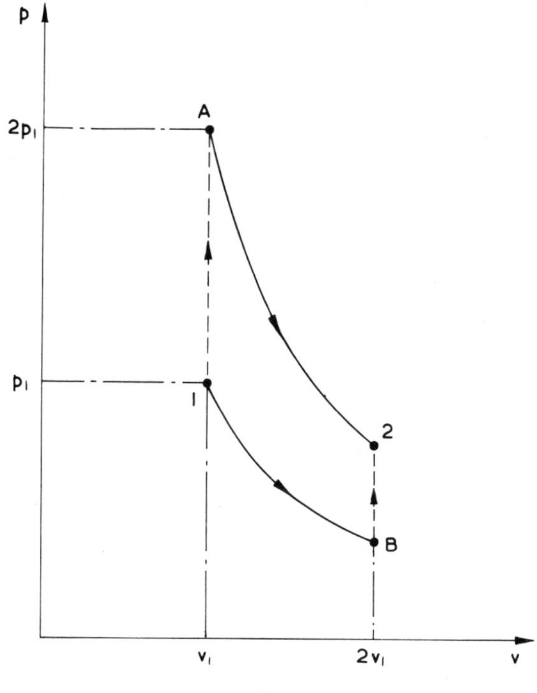

Fig. P.1

(a) *Process 1–A–2 in Fig. P.1.* In this process, the pressure is first raised by driving the paddle wheel while the piston is held stationary. When all eddies have subsided, the air settles to Stable State A at a pressure $2p_1$. The piston is then allowed to move in a fully resisted manner until the volume of the air is $2v_1$, process A–2 thus being isentropic.

(b) *Process 1–B–2 in Fig. P.1.* In this process, the air is first expanded isentropically to B and rotation of the paddle wheel while the piston is held stationary then brings the air finally to Stable State 2.

Calculate the work quantities in each of the four subprocesses 1—A, A—2 and 1—B, B—2. Thence confirm that, in accordance with Corollary 1 of Section 4.2, the net work output is the same in these two alternative adiabatic processes between Stable States 1 and 2.

Calculate the change of entropy in each of the two processes 1—A and B—2, and sketch the temperature—entropy diagram for each of the two overall processes.

Answer:

$$W_{1-A} = - \frac{p_1 v_1}{\gamma-1}, \qquad W_{A-2} = \frac{2p_1 v_1}{\gamma-1}\left(1 - \frac{1}{2^{\gamma-1}}\right),$$

$$W_{net} = \frac{p_1 v_1}{\gamma-1}\left(1 - 2^{(2-\gamma)}\right).$$

$$W_{1-B} = \frac{p_1 v_1}{\gamma-1}\left(1 - \frac{1}{2^{\gamma-1}}\right), \qquad W_{B-2} = - \frac{1}{2^{\gamma-1}}\left(\frac{p_1 v_1}{\gamma-1}\right),$$

$$W_{net} = \frac{p_1 v_1}{\gamma-1}\left(1 - 2^{(2-\gamma)}\right)$$

$$\Delta S_{1-A} = \Delta S_{B-2} = c_v \ln 2.$$

Answers to Problems 3.1, 3.2, 5.1, 6.1, 7.1 and 11.5

3.1 (a): (1) +, (2) +, (3) +.

(b): (1) −, (2) +, (3) 0.

3.2 (a) 0; (b) 0; (c.1) − , (c.2) 0; (d) 0; (e) + .

5.1

		(1)	(2)	(3)
3.1 (a)	ΔE	−	−	−
	ΔU	0	0	0
3.1 (b)	ΔE	+	−	0
	ΔU	+	0	+

		(a)	(b)	(c)		(d)	(e)
				(1)	(2)		
3.2	ΔE	0	0	+	0	0	−
	ΔU	0	−	−	−	+	0

6.1

	(a)	(b)(1)	(b)(2)	(c)	(d)	(e) Missile	(e) Air	(e) Both	(f) Accumulator	(f) Resistor	(f) Both
Q	0	0	0	0	−	0	0	0	0	−	−
ΔE	+	0	0	0	−	−	+	0	−	0	−
ΔU	+	0	+	0	−	0	+	+	−	0	−

7.1

Process	System	Q	W	ΔE	ΔU	$\Delta(PE)$
First	(a)	0	0	0	+	−
First	(b)	0	+	−	+	−
First	(c)	0	−	+	0	+
Second	(a)	0	−	+	0	+
Second	(b)	0	−	+	0	+
Second	(c)	0	+	−	0	−

11.5

Case	1	2	3	4	5	6
Source temperature, °C		−6.3				
Sink temperature, °C						−
Heat from source, kJ/h			−	0		
Heat to sink, kJ/h		9 000	−		11 000	
Power output, kW	1.25	− 0.556			−1.39	
Thermal efficiency, %	45	−		−	−	−
Coefficient of performance	−		−	−		−
Reversible, irreversible, or impossible?	Irrev.	Rev.	Imposs.	Irrev.	Rev.	Imposs.
CHPP, refrigerator, or heat pump?	CHPP	Refrig.	−	−	Heat pump	−

Index

triple-product theorem, 277, 284

universal (molar) gas constant, 226
unrestricted equilibrium, 186, 371
unstable equilibrium, 12

van't Hoff Equation, 366
van't Hoff Equilibrium Box, 366, 368
velocity coefficient, 147
volume fraction, 228
volumetric analysis of gas mixture, 228

work, 6, 28
    displacement, 30, 68, 184

electrical, 31, 265, 384
external, 105, 113, 185
gross, 101, 181, 182, 187, 188,
    211, 213, 218, 221, 222
indicated, 31
internal, 105, 112, 131, 143, 184,
    216, 218, 221
mechanical, 6, 29
shaft,
    non-flow process, 185−189
    steady-flow process, 66, 68, 136,
        187, 190−193, 370
work equivalent, 378
work interaction, 5, 7, 10, 28, 33, 57,
    59